Introduction to Relativity Volume I

$E = mc^2$ is known as the most famous, but least understood equation in physics. This two-volume textbook illuminates this equation and much more through clear and detailed explanations, new demonstrations, a more physical approach and a deep analysis of the concepts and postulates of Relativity.

The first part of Volume I contains the whole Special Relativity theory with rigorous and complete demonstrations. The second part presents the main principles of General Relativity, including detailed explanations of the bending of light in the neighborhood of great masses, the gravitational time dilatation and the principles leading to the famous equation of General Relativity: $D(g) = \kappa.T$. The most important cosmological predictions are then described: the Big Bang theory, black holes and gravitational waves. Plentiful historical information is contained throughout the book, particularly in an ending chapter depicting the scientific and epistemological revolution brought about by the theory of Relativity.

Both volumes place an emphasis on the physical aspects of Relativity to aid the reader's understanding and contain numerous questions and problems (147 in total). Solutions are given in a highly detailed manner to provide the maximum benefit to students.

Introduction to Relativity
Volume I
In-Depth and Accessible

Paul Bruma

CRC Press
Taylor & Francis Group
Boca Raton London New York

CRC Press is an imprint of the
Taylor & Francis Group, an **informa** business

First edition published 2022
by CRC Press
6000 Broken Sound Parkway NW, Suite 300, Boca Raton, FL 33487-2742

and by CRC Press
4 Park Square, Milton Park, Abingdon, Oxon, OX14 4RN

CRC Press is an imprint of Taylor & Francis Group, LLC

© 2023 Paul Bruma

Library of Congress Cataloging-in-Publication Data
Names: Bruma, Paul, author.
Title: Introduction to relativity : in-depth and accessible / Paul Bruma.
Description: First edition. | Boca Raton : CRC Press, 2023. |
Identifiers: LCCN 2022007077 (print) | LCCN 2022007078 (ebook) |
ISBN 9781032056746 (v. 1 ; hardback) | ISBN 9781032062433 (v. 1 ; paperback)
| ISBN 9781032056760 (v. 2 ; hardback) | ISBN 9781032062440 (v. 2 ;
paperback) | ISBN 9781003201335 (v. 1 ; ebook) | ISBN 9781003201359 (v. 2 ; ebook)
Subjects: LCSH: Relativity (Physics)
Classification: LCC QC173.55 .B75 2023 (print) | LCC QC173.55 (ebook) |
DDC 530.11—dc23/eng20220716
LC record available at https://lccn.loc.gov/2022007077
LC ebook record available at https://lccn.loc.gov/2022007078

ISBN: 9781032056746 (hbk)
ISBN: 9781032062433 (pbk)
ISBN: 9781003201335 (ebk)

DOI: 10.1201/9781003201335

Typeset in Palatino
by codeMantra

It is hard, for anyone who really understands it, to resist the charm of this theory.

Albert Einstein about General Relativity.

May this book let the reader share this feeling, and for Special Relativity as well.

Contents

Preface by Jean Iliopoulos, Dirac Prize 2007

The beginning of the last century witnessed two major revolutions in the physical sciences which changed profoundly our perception of space and time and our views on the structure of matter. They are known as the theory of Relativity, for the first, and quantum mechanics, for the second. They both have a well-deserved reputation of complexity, and this has scared people away. Only specialized professionals dared to study and understand them. Yet, both are omnipresent in our everyday lives. The tiny chips which form the heart of our smartphones could not be designed without quantum mechanics, and our GPS would be grossly inaccurate without the theory of Relativity. Maybe it is time to try to demystify these theories and make them accessible to those among the younger generation who are interested, well motivated, and have followed a good science course at high school.

This book tries to meet this challenge for the theory of Relativity. By "demystifying" I do not mean a kind of vulgarization which avoids the difficulties by offering more or less convincing plausibility arguments. The ambition of this book is to explain every point by presenting a rigorous and complete derivation. It starts from first principles, sets the axioms and develops the theory step by step in a fully deductive way. Every assumption is solidly anchored in experimental results.

The book contains two volumes. In the first, we find the general principles and the main results, first for the Special theory (Chapters 1 to 5) and then for the General theory of Relativity (Chapters 6 and 7). More advanced points are left for the second volume. The book is not "easy", and the reader should follow every step carefully and repeat the calculations, but the end result is highly rewarding. At the end he, or she, will have a very good working knowledge of this very beautiful theory. A great help is provided by well-chosen problems and exercises, which we find at the end of each chapter. I strongly advise the reader to try hard to solve them. The solutions can be found at the end of each volume, but it would be a mistake to look there directly for the answers.

There exist many excellent books on the theory of Relativity, but, contrary to most of them, the author of this one does not require the reader to have a background in physics and mathematics beyond what one can reasonably expect from a good high school graduate. The result is amazing. Incredible as it may sound, the author wins his bet. He shows that the theory of Relativity is not "difficult". It is fully accessible to the kind of readership he set to meet. It is the book I wish I had 65 years ago, when I finished high school.

J. ILIOPOULOS
Director of Research Emeritus
Ecole Normale Supérieure
Paris

Acknowledgments

Christine Bruma: I would like first to acknowledge my wife Christine for her continuous encouragement and unlimited support.

Ryan Rohm, Professor of Physics and researcher: Many thanks for all your very valuable remarks, comments and suggestions in the course of your detailed review.

Jean Iliopoulos, Dirac Prize 2007, member of the French Academy of Sciences: Many thanks for your interesting remarks and suggestions targeted at making the book even more attractive.

Yves Meyer, Abel Prize 2017 in mathematics: Many thanks for your advices and your positive feedback on the pedagogical character of this book.

Dominique Delande, Research Director in physics at CNRS: Many thanks for your very interesting physical remarks and comments.

Sophie Cribier, Professor of Physics, Paris University: Many thanks for your very useful remarks, especially in order to improve the pedagogical aspects.

Hugon Karwowski, Professor Emeritus of Physics, University of North Carolina: Many thanks for your very useful remarks.

François Saint-Jalm, Professor of Physics: Many thanks also for your very accurate physical remarks and comments.

Thomas Cerbu, Professor of Literature at University of Georgia: Many thanks for your review and your advices.

Gérard Noblins, Engineer: Many thanks to my very dear old friend who wanted this document to be a jewel, containing all crystal clear explanations, and who kept on challenging each and every point, even those which seem natural.

Loïc Melscoët, Engineer: Many thanks to my very dear old friend who gave me the motivation to undertake this book, sharing his passion for physics and, in particular, the different facets of Relativity.

Michel Mouly, Engineer: Many thanks for not counting your time in the numerous and always very interesting discussions we had, and where you shared your very high expertise in physics, and your great talent in explaining even complex matters. This book would not have been possible without you.

Jean Bruma, Engineer: Many thanks to my dear brother who kept on wanting simpler and simpler explanations.

Monique Mémet, Professor of English: Many thanks to my very dear sister-in-law for your great help in improving the English used in this document.

Danièle Mémet, Graphist, head of the company Graphir®: Many thanks to my very dear sister-in-law who contributed in the design of the cover of this book.

Daniel Kolski, Engineer: Many thanks for your help and good suggestions with many good ideas.

Eric Gradeler, Engineer: Many thanks to my very dear old friend who issued many judicious comments and remarks which very significantly improved the book.

Gregory Colton: Many thanks for your interest in this book and your numerous comments that improved the clarity and the English used.

Régine Hollander, Professor of English: Many thanks for your contribution in improving the English used in this document.

Alois Cerbu, Science student at Berkeley: Many thanks for your advices and your contribution in improving the English used in this document.

Author

Paul Bruma is a French engineer who graduated from Institut Polytechnique de Paris, Telecom-Paris This curriculum includes a broad program in mathematics and physics, equivalent to a master's degree in science.

After a career with the international telecom equipment manufacturer Alcatel-Lucent (former Bell Labs), Paul Bruma resumed physics studies, which was his favorite discipline as a student. Regarding Relativity, Paul Bruma found that most text books lack explanations and are very mathematically oriented. This explains why this subject appears quite complicated to students, and many frequently switch from one book to another in search of the missing explanations in their text books. Relativity being an essential subject, Paul Bruma took up the challenge of making a book that contains all explanations and in a manner which is as accessible as possible while always being absolutely rigorous. This induced him to adopt a more physical approach than most authors.

In his previous career in the high-tech industry, Paul Bruma had many opportunities to write technical documents explaining complex subjects and to train teams. In this domain, if technical specifications are not written in a clear, detailed, step-by-step and unambiguous manner, the implementation teams won't work effectively and the outcome will likely differ from what was intended. In contrast, academic authors often consider that students should fill by themselves some missing steps or explanations as part of the pedagogical process. Paul Bruma believes that this method is not the best one for Relativity because it is a domain where common sense often misleads.

General Introduction

At the end of the 19th century, science had achieved so many important accomplishments that it was commonly expected to explain all in the near future. There were yet some open issues, but these were generally considered as minor, in particular the one raised by Maxwell's theory of electromagnetism.

Trying to solve this problem, a young engineer named Albert Einstein, aged 26, totally unknown by the scientific community, showed in 1905 that time and space were not absolute, contradicting a fundamental and well-established Newtonian paradigm. The same year, he presented the main new laws of dynamics, including the relation $E = mc^2$, which later became the most famous of physics.

The impact of his new theory reached all of physics and even regions of philosophy: it meant that almost all our physical laws were wrong, which could not be noticed because our instruments were not precise enough. Besides, we could not witness phenomena involving extremely very high speeds where our laws were very wrong.

The next important step occurred in 1908 and was due to H. Minkowski who showed that our universe is a 4D space-time continuum. He also presented the new concepts and mathematical tools which were adapted to this new reality, and which proved to be extremely useful until today.

A final huge step occurred in 1915, again due to Einstein when he issued the General Relativity theory. He showed that gravitation does not exist, but its effects stem from accelerations generated by space-time distortions of our universe due to the presence of masses. J.J. Thomson, president of the British Royal Society, described this theory as "one of the greatest, if not the greatest achievement in the history of human thought".

**

For the first time in history, common sense and intuition play a counterproductive role in Relativity. Therefore, we will start with an in-depth presentation of the concepts and postulates of Relativity; then from this solid basis, the laws of relativity will be demonstrated with the rigor and logic of mathematics. In order to facilitate understanding, a very gradually incremental approach will be used, along with many concrete scenarios and detailed explanations. Last but not least, historical aspects will also be carefully depicted.

Both volumes comprise the whole theory of Special Relativity and an introduction to General Relativity and cosmology. In Volume I, the chosen demonstrations emphasize the physical aspects and require a lower mathematical background than in Volume II. The Volume II goes more in depth, contains complementary explanations and case studies and addresses additional subjects especially in General Relativity and cosmology. The more complex mathematical means used in Volume II are carefully explained for readers who are not familiar with them.

1

Relativity of Time and Distance

Introduction

This chapter presents the revolutionary steps taken by Einstein to solve the problems raised by Maxwell equations at the end of the 19th century, especially the disconcerting result that the speed of light must be independent of the speed of its source. Einstein stated that we must accept this result, and showed that if it appears absurd, it is because our basic conceptions of the absolute time and distance were wrong, and so was our ether assumption. His reasons will be explained, together with his strong belief in Galileo's principle of Relativity. Hence, the basic notions of time and distance will be carefully examined within a given inertial frame in order to know exactly what remains valid and on what basis. The main postulates of Special Relativity will thus be precisely presented. Then, when addressing the general case of objects in different inertial frames, Minkowski's new and powerful concepts of "event", "proper time" and "proper distance" will be carefully presented, all the more as they will be very useful in both Special and General Relativity.

1.1 Time and Space Are Not Absolute

The whole Special Relativity was developed by Einstein using only one additional and disconcerting postulate, the *light speed invariance postulate*, while keeping all other postulates which were compatible with this new one. Hence, let's examine this new postulate, and we will see that its direct consequence is no less than the Relativity of time and distance:

1.1.1 Light Speed Invariance: An Affront to Common Sense

The light speed invariance postulate states that **the speed of light is the same in all inertial frames, independently of the speed of the source which has emitted it**. The last part is counterintuitive, and even an "affront to common

DOI: 10.1201/9781003201335-1

sense" (an expression used by *The Times of London*[1]). Let's illustrate this affront with a simple experiment:

> Imagine that you stand on Earth while your friend flies a rocket moving away from Earth at the speed of 0.8 times the speed of light: 0.8 c. You send a light pulse in the direction of the rocket, and you ask your friend to measure the speed at which he sees this light pulse passing. The classical reasoning, and even our common sense, say that this speed is: c − 0.8c = 0.2 c, but actually, your friend measures the pulse going at the speed c, which is disconcerting.
>
> Einstein answered that the reason why we find this disconcerting is that our intuition and our common sense stem from our experience, but we never witnessed speeds in the range of the speed of light.

Note that his postulate concerns inertial frames. As a first definition, an inertial frame is a frame which is fixed to a rigid object that is submitted to no force, and thus incurs no acceleration. Inertial frames play an important role in Special Relativity; hence, this concept will be precisely defined in Section 1.2.1.

Before seeing the drastic consequences of this new postulate, let's figure out what light speed represents, considering that light speed plays a special role in Relativity:

Light propagates in the vacuum at the speed of c = 299,792 km/s. For reference, this equals roughly 3×10^8 **m/s**, or 1.079 billion km/h, or 671 million miles/h. Light thus appears to be enormously fast in the context of our everyday activities: still today, light speed is 200,000 times faster than our fastest moving objects (bullets). However, there is one case where light speed can be felt: when you phone a person on a different continent, you can perceive a delay in the voice response. This is due to the transmission by satellite, whose very long distance (36,000 km) induces a delay to the electromagnetic signal's travel in the range of 400 ms.

Conversely, when we consider the cosmos, light speed has very significant effects: for instance, if a light pulse from Earth is sent out to the cosmos, it will take:

- 1.3 seconds to reach the Moon.
- 8.1 minutes to reach the Sun.
- 40 minutes to reach Jupiter.
- 4.2 years to reach our nearest star, Proxima Centauri.
- 100,000 years to traverse our galaxy.
- 2.55 million years to reach our nearest galaxy, Andromeda.

Light from the furthest known galaxy traveled for 13.2 billion years to reach us, and so was emitted only 500 million years after the Big Bang.

Some history: It is in 1676 that humanity discovered that the speed of light was finite, and this was due to the Danish astronomer, Ole Rømer. He successfully predicted the 10-minute delay in the occurrence of an eclipse of a Moon of Jupiter, and explained this delay by the fact that Jupiter was at a longer distance from Earth than during the previous observation; hence, light needed more time to reach us. His estimation for light speed was beyond 200,000 km/s. Before that, some famous philosophers and scientists believed light speed was infinite (Aristotle, Kepler, Descartes, etc.), but others finite (Empedocles, Newton, Bacon, Fermat, etc.). In 1849, the French scientists H. Fizeau and L. Foucault made some experiments which determined the correct value of light speed, with an accuracy better than 5%.

1.1.2 Time and Distance Relativity

Einstein's famous train scenario will show us that the invariance of light speed implies the Relativity of time and distance:

1.1.2.1 Einstein's Famous Train Scenario

A train is moving with a constant velocity v and passes by a railway station. On its platform, two points are marked, A and B; whenever B lines up with the front end of the train, A lines up with its rear. Then, at the precise moment when the front end of the train is in front of B, each point (A and B) simultaneously emits a pulse of light. One passenger is sitting exactly in the middle of the train, at the point M in Figure 1.1. *Question: Will he receive both pulses simultaneously?*

Observers on the ground will say that the passenger in the middle of the train, M, will receive the signal from B before that from A, because during the time taken by the light pulses to reach him, the train will have advanced toward B. In Figure 1.1, the passenger in the middle of the train will be at the point M' when receiving the first pulse, that is the one emitted from B since it has a shorter distance to cover than the one from A.

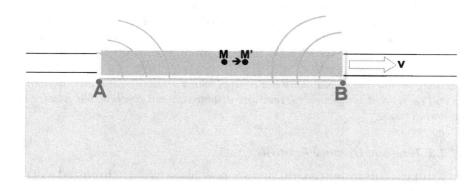

FIGURE 1.1
Einstein's famous train scenario.

Let's consider the same scenario, but from passengers inside the train: they see two signals emitted from both ends of the train, and propagating at equal speed, c, in accordance with the light speed invariance postulate. Hence, these passengers will say that these two pulses will reach the middle M of the train simultaneously since they will cover equal distances at the same speed.

These statements from the ground and from the train are contradictory; who is right?

Looking with a critical mind at each step of the above reasoning, we realize that we made an implicit assumption. Indeed, when we considered the scenario from the train perspective, we implicitly assumed that passengers saw that both pulses were *simultaneously* emitted, because they were simultaneously emitted from the platform as stated in the scenario set out.

Conversely, when we consider this scenario from the ground perspective, we did not make any implicit assumption; hence, the reasoning made by the observer on the ground can only be correct.

This scenario thus shows that *simultaneity* is relative to an inertial frame. Two distant events (e.g., the emission of the pulses from A and B) which are simultaneous in one inertial frame K (the platform) are not considered simultaneous in other inertial frames K' (e.g., the train) moving relative to the frame K.

Remark 1: In classical physics, we do not have this absurd situation because the passengers inside the train see the signal from B arriving with speed $(c + v)$ and the signal from A arriving with speed $(c - v)$. Consequently, they see the pulse from B arriving first, like observers on the ground. It is the light speed invariance postulate which gives rise to the Relativity of simultaneity. Conversely, if light speed were infinite, we would not have this absurd situation.

Remark 2: An experiment can be made, placing a camera on the ground, and another one in the train. Both cameras will confirm that the passenger in the middle of the train will first receive the signal from the front, and then the one from the back. However, the time difference between the two signals is hardly detectable.

Complements on Einstein's famous train scenario are with Exercise 2.11, and in Volume II Section 1.3.1. In particular, a calculation is made with a distance AB of 200 m and a train speed of 300 km/h. The result is that the time difference considered from the train between the two light pulses emissions is: 1.8×10^{-13} seconds, which is extremely small, but can be detected with today's atomic clocks.

Having shown that simultaneity is a concept which is relative to an inertial frame, we will see its first consequences: time and distance are not absolute, but relative to an inertial frame.

1.1.2.2 Time and Distance Relativity

Let's still use this train scenario, and we now place two synchronized clocks on the platform, one at the point A and the other one at the point B. These clocks, being synchronized display the same time for any observer on the platform. However, passengers inside the train do not see these two clocks showing the same time because of the Relativity of simultaneity, as previously shown. Hence, the time on the platform is not the same as the time in the train, meaning that Time is relative to a given inertial frame.

This same scenario can be analyzed symmetrically by considering the train being fixed while the ground moves (in the opposite direction). The same reasoning leads to the same conclusion: if synchronized clocks are placed at both ends of the train, these clocks will not show identical times to observers on the ground. This again means that Time is relative to an inertial frame.

The simultaneity of events being relative to an inertial frame, the very notions present, past and future are also not absolute, but relative to a given inertial frame, and this is a serious restriction to the notion of time.

Moreover, time Relativity implies distance Relativity: the length of any segment AB can indeed be defined as the time duration required for light to travel from A to B, multiplied by c (light speed). Since time is relative to an inertial frame, while the speed c is not, distance must also be relative to an inertial frame.

Furthermore, this train scenario shows that time and space are linked together: in the train, let's place a clock at every window; we make sure that these clocks are synchronized, meaning that they all show the same time for all passengers inside the train. Due to the Relativity of simultaneity, observers on the platform see each clock showing a different time. This means that the location of the clock inside the train has an influence on the time of this clock seen from the ground: time and space are linked together.

Time and space Relativity were revolutionary as they ran contrary to Newton's assertion of the absolute frame. According to Newton, there is an absolute rest frame; it has always existed and remained unchanged. It even pre-existed before any matter, without any relation with anything. The time given by this absolute reference frame is the same for all, independently of the positions and the motions of the observer and the observed clock, and the same for the distance. However, there were a few early opponents, in particular Berkeley and Leibnitz who stressed that it must be possible to check any physical statement, and prophetically stated that time and space were properties of matter. Nevertheless, there was some strong evidence in favor of Newton, in particular the existence of fixed visible stars.[2] Newton was considered the greatest scientist ever, and as there was no evidence against his assertion, the scientific community adopted it. One century later, the ether hypothesis came reinforcing Newton's absolute frame, as we will further see. At the time of Einstein, there were very few opponents to the absolute frame, but notably Ernest Mach.[3]

1.2 What Led Einstein to His Conclusion and the Principle of Relativity

Electromagnetism was the first domain in physics involving extremely high speeds. When J.C. Maxwell issued his famous equations in 1862, these raised problems with respect to the principle of Relativity, and even to the common sense as we will see. Maxwell's equations are rather complex and not necessary to understand well the theory of Relativity.[4] Hence, we will only present hereafter the issues they raised and then the way Einstein solved them. For this, we first need to carefully present the principle of Relativity, all the more as it is key in Relativity.

1.2.1 The Principle of Relativity

The principle of Relativity, later called the inertial frames equivalence postulate, states that **all natural phenomena can be described according to the same physical laws in all inertial frames**. Its author, Galileo, noticed that if you are in a boat moving at constant velocity, everything behaves as if the boat were fixed. He had a good illustration: "butterflies flutter like on the ground". Einstein also had a good illustration: if you are sitting inside a train moving at constant velocity and with no windows, there is no experiment that allows you to know the train speed, nor even whether it is moving.

This principle applies to inertial frames; hence, let's first see the definition of an inertial frame:

1.2.1.1 Inertial Frame Definition

The concept of inertial frame is based on the **law of inertia**, which states: **a body which is not submitted to any force continues in a state of rest or in uniform motion along a straight line**. Such a body is called inertial; then an inertial frame is a frame which is fixed with this inertial body (attached to it). The three axes and the origin of this inertial reference frame can be chosen arbitrarily. We will generally use Cartesian orthonormal frames of reference because they are simpler to use. The notions of distance and time will be further examined in the next chapter, but for the time being, their intuitive definitions suffice.

Let's have a first inertial frame K fixed with an inertial object A. Any object B moving with a constant velocity $\vec{v}$ relative to K is also inertial, and this is consistent with the famous second Newtonian law: $\vec{F} = m\vec{a}$, where $\vec{a}$ is the acceleration relative to the inertial frame K. The object B is thus subjected to no force; hence, it is inertial and any frame K' fixed with the object B is also an inertial frame.

> Conversely, a frame K" that is not moving with a constant velocity relative to an inertial frame is not inertial. From a mathematical perspective, the set of inertial frames constitutes an equivalence class.

Historically, Galileo believed that there existed one special frame: the one which was fixed relative to the stars that were considered fixed. Galileo then defined the concept of inertial frame, also called Galilean frame, as a frame which is in uniform motion relative to this fixed frame. Newton supported this vision, which matched with the absolute frame of his theory.

However, we now know that perfect inertial frames do not exist: our Earth is not fixed nor in uniform motion since it turns around itself and around the

Sun. The Sun is also orbiting within our galaxy, and our galaxy too is moving very fast: there is no fixed star. The problem with such imperfect inertial frames is that their accelerations generate inertial forces. As an example, when you are in a lift starting to move downward, you feel a force upward.[5] Nevertheless, the approximations resulting from considering the Earth frame as inertial are acceptable in many instances.

Brief history of the principle of Relativity:

The first revolutionary step was due to the Polish astronomer Copernicus in early 16th: the Earth is not the center of the universe around which the Sun and the planets are turning, but the opposite is true! He found that the heliocentric scheme brings more simplicity and harmony in the descriptions and calculations of planets' trajectories. One century later, Galileo confirmed Copernicus's heretical statement after having observed that Jupiter also has Moons.

Moreover, Galileo issued the famous law of inertia, which went contrary to the dominant thinking inherited from Aristotle who stated: "everything that is in motion must be moved by something". This vision was supported by experience: if you let any object go, its speed will decrease and the object will stop. People did not imagine that this deceleration was due to friction with the air and/or the soil. Conversely, when astronomers could observe stellar objects, they could see objects in motion where there was no friction at all, and indeed those which were not orbiting had perfect inertial trajectories. Galileo stated that this is also true on Earth, but we did not notice it because of friction. He then issued the principle of Relativity after noticing that inside a boat in uniform motion, everything behaves like on the soil: "Inertial motion is like nothing".

1.2.1.2 The Galilean Coordinate Transformation and the Classical Velocity Additive Law

Consider two inertial frames K and K′, with K′ moving with the constant velocity V relative to K along the OX axis. We also say that K′ is obtained from K by performing a *boost*. We choose that when the origins O and O′ are at the same point, the times of K and K′ are null: T = T′ = 0. In other words, K and K′ have the same origin-event. Besides, K and K′ have the same length unit.

At a random time t, a point object M is seen in the frame K with its coordinates being (x, y, z). Time being absolute, the time t is the same in both frames, so at the same time t in K′, this same object is seen with the coordinates (x′, y′, z′). The relation giving (x′, y′, z′) knowing (x, y, z) is called the Galilean coordinates transformation, and we will express it with the support of Figure 1.2:

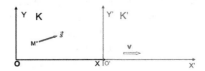

FIGURE 1.2
Galilean velocity addition law.

At the time t, we have: $\overline{O'M} = \overline{O'O} + \overline{OM}$, meaning:
$$(x', y', z') = (-v.t, 0, 0) + (x, y, z).$$
Thus: $x' = x - vt$, $y' = y$ and $z' = z$, which is the Galilean coordinates transformation law.

We then want to find the relation between the velocity $\vec{S}$ of the object M in K, and its velocity $\vec{S'}$ in K'. Let's make the derivative relative to the time of the previous relations:
$$\frac{dx'}{dt} = \frac{dx}{dt} - v; \quad \frac{dy'}{dt} = \frac{dy}{dt}; \quad \frac{dy'}{dt} = \frac{dy}{dt}, \text{ meaning: } \vec{s'} = \vec{s} - \vec{v}, \text{ which is the clas-}$$
sical velocity additive law.

1.2.1.3 The Principle of Relativity Was Supported by the Famous Newtonian Laws

1.2.1.3.1 *The 2nd Newtonian Law Supported the Principle of Relativity*

Imagine that our object M has an inertial mass m. For the sake of simplicity, we now set its speed $\vec{S}$ to be parallel to $\vec{v}$ and to both OX and O'X' axes. This object is seen in K having an acceleration $\vec{a} = \dfrac{d\vec{s}}{dt}$. The 2nd Newtonian law says that this acceleration is due to a force $\vec{f}$ incurred by this object, with $\vec{f} = m\,\vec{a}$.

In another inertial frame K', the principle of Relativity implies that: $\vec{f'} = m'\,\vec{a'}$ since this postulate states that all physical laws are the same in all inertial frames. We will see that this relation is respected by the Newtonian force:

We saw that: $\vec{s'} = \vec{s} - \vec{v}$; hence: $\vec{a'} = \vec{a} - \dfrac{d\vec{v}}{dt} = \vec{a}$ ($\vec{v}$ being constant since K and K' are inertial). Besides in classical physics, the mass was assumed to be constant and invariant, meaning identical in all inertial frames: $m = m'$. Consequently, $\vec{f'} = m'\,\vec{a'} = m\,\vec{a} = \vec{f}$, which means that the Newtonian force is the same in all inertial frames. This identity was confirmed by all experiments; hence, the Newtonian force strongly supported the principle of Relativity.

1.2.1.3.2 *The Newtonian Gravitational Force also Supported the Principle of Relativity*

The classical force of gravity was $\vec{W} = m_g\vec{g}$, where m_g is the object gravitational mass and $\vec{g}$ a vector oriented toward the center of the great mass, proportional to its gravitational mass M and inversely proportional to the square of the distance to the center of this mass.

In another inertial frame K', the principle of Relativity implies that the force of gravity must be: $\vec{W'} = m_g\,\vec{g'}$. At the same point, the vector $\vec{g}$ is the same whatever the inertial frame. Furthermore, the gravitational mass is equal to the inertial mass[6], and thus invariant (identical in all inertial frames). Hence, the gravitational force must be the same in all inertial frames, which was confirmed by all experiments: for example, if you weigh yourself with a scale, you will find the same value whether you are on the soil, or inside a boat moving with constant velocity, or inside a lift also moving with constant velocity (after the acceleration phase).

1.2.2 The Issues Raised by Maxwell's Equations

For the first time in 1865, the principle of Relativity did not apply to a physical law: Maxwell's equations. The following problems illustrate this discrepancy:

1.2.2.1 Case of an Electrical Charge Moving Beside a Magnet

If an electric charge is moving with constant velocity beside a magnet, Maxwell's equations enable us to calculate the electric and magnetic fields incurred by this charge, and consequently the force that it incurs. If we consider this scenario from the inertial frame where the magnet is fixed, Maxwell's equations say that the charge incurs a magnetic field and no electric field. If we consider this same scenario from the inertial frame where the charge is fixed and the magnet moving, Maxwell's equations tell us that the charge incurs not only a magnetic field, but also an electric field. Even if the resulting forces were the same, Einstein and most scientists considered that this was against the principle of Relativity: the same physical phenomenon was indeed explained by different laws in different inertial frames.

1.2.2.2 The Galilean Coordinates Transformation Law is Contradicted

If we consider Maxwell's equations in an inertial frame K′ moving at the speed v relative to another inertial frame K, we can use the Galilean coordinate transformation law to replace (x′, y′, z′, t) in these equations with (x − vt, y, z, t). However, we are surprised to obtain equations that do not have the same form as Maxwell's equations in K. This again contradicts the principle of Relativity.

1.2.2.3 The Light Speed Invariance Issue

Maxwell's equations show that electromagnetic waves propagate at a speed which is constant and equal to $c = 1/\sqrt{\varepsilon\mu}$ where ε and μ, respectively, are Coulomb's and Faraday's constants. The value of c could be determined since many experiments enabled us to know the values of ε and μ. The result was a value of c very close to the speed of light that was known; hence, Maxwell deduced that light was an electromagnetic wave. However, a problem arose since ε and μ were found to be identical in different inertial frames, independent of their motions; this implied that the speed of light is the same in all inertial frames, independent of the speed of the source that has emitted it: "an affront to common sense!"

However, this problem was masked because the vast majority of scientists believed in the existence of the ether. Indeed, the scientist A. Fresnel stated in the early 19th that light propagation was due to the vibrations of a luminiferous ether, which was supposed to be a very light substance filling the universe. Light propagation was thus due to a similar mechanism as sound: like sound, light would generate pressure variations in an elastic media, and these variations would then propagate at a given speed which is specific to this media.

This ether assumption masked the light speed invariance issue because the ether was logically assumed to be static within the Newtonian absolute reference frame; hence, the speed of light given by Maxwell's equations was c only in this absolute frame. In other inertial frames, there would be an "ether wind" effect which would reduce or increase the speed of light depending on whether light propagates in the direction of the ether or against (similarly as sound propagates more slowly if it goes against the wind).

Fresnel previously made important findings; hence, the scientific community adopted his idea which matched well with the Newtonian absolute reference frame. There were, however, some isolated opponents pointing out the total absence of evidence of such ether, in particular E. Mach again.

Scientists were eager to characterize this luminiferous ether. Hence, Michelson-Morley made their famous experiment (1881–1887) aiming at measuring the "ether wind" effect. This experiment is detailed in Volume II Section 1.1.2, and can be summarized as follows: a light source is split into two beams which follow two perpendicular paths before recombining and interfering. Being perpendicular, these two beams have different directions relative to the ether motion, and consequently incur differently the "ether wind" effects. When these beams recombine, this difference results in a phase difference, which can be detected with the interference pattern. This experiment was carried out many times and along all directions, and the results were always the same: no "ether wind" was ever detected, meaning that the ether has no influence on light propagation.

In case of ether nonexistence (or absence of an ether wind effect), Maxwell's equations raised the disconcerting light speed issue of its independence from the velocity of the source which has emitted it. It is remarkable that even after the Michelson-Morley experiment, most scientists continued to believe in the ether existence.

1.2.3 Einstein's Solution to These Issues

Einstein was extremely audacious to reject not only the Newtonian absolute time and space, but also the ether existence. Einstein was already skeptical about the ether existence because of the stellar aberrations[7] and Fizeau's measurements of the speed of light in moving water. These observations and experiments are described in Volume II Sections 1.1.1 and 1.5.1, and they showed that the ether could not be fixed relative to the Earth: the ether could not be dragged by the atmosphere of our planet like the air (or very partially), nor by moving water.

Besides, Einstein had a strong philosophical belief in the principle of Relativity since inertial frames are symmetrically related[8]. He added to this principle that no frame should be privileged, which contradicted Newton's absolute frame. Moreover, he stated that this principle must be respected by all physical laws, whereas it was previously considered as a guideline.

Einstein imagined the case of an observer moving at light speed and following an electromagnetic wave: he would logically see a "spatially oscillatory electromagnetic field at rest", but this was against Maxwell equations. Moreover, if this observer emits a light pulse, it would go with the speed c relative to him, but also to the frame where the first wave was emitted, which is absurd. However, Einstein noticed that this was not contradictory with the principle of Relativity, and realized that it was absurd because we used the classical Galilean additive velocity law. Einstein understood that this law was wrong since it was based on the absolute character of time and space, and subsequently found how time and distance are transformed from one inertial frame to another one: by the "Lorentz transformation", a new formula that had been found few months before by the French scientist H. Poincaré[9].

Using this transformation, Maxwell's equations keep the same form in all inertial frames. Einstein could then deduce the new velocity *composition* law, which replaced the classical velocity *addition* law, and this new law showed the existence of one invariant speed: the speed of light. The "affront to common sense" was solved!

In another step the same year, he showed that the inertial mass of an object increases with its speed and tends to the infinite as the object speed tends to c, which was also revolutionary. Therefrom, he deduced the impossibility to reach the speed of light, so that the absurd phenomenon he imagined with an observer traveling at the speed c could not happen in reality: it thus all came full circle.

This had enormous consequences: all our laws of dynamics must be wrong since they respect the principle of Relativity while using the (wrong) classical Galilean coordinate transformation law instead of the Lorentz transformation. However, when objects' speeds are much smaller than c, the differences between the results of the (wrong) classical laws and the (correct) relativistic ones are extremely small, which explains why the former were thought absolutely correct. Conversely, for extremely high speeds, the differences are huge. Three months later, Einstein published his famous equation, $E = mc^2$, showing that the mass is the measure of the energy contained in an object.

The Relativity revolution involves a fundamental change in the notions of time and space on which all of physics is based. Hence, before demonstrating the new relativistic laws, we are invited to carefully examine these notions.

1.3 Examination of the Notions of Distance and Time

We will first examine the notions of distance and time within one inertial frame on the basis of the knowledge of Einstein's time. We will then address the general case of moving inertial frames, and see the new key concepts which are appropriate in this context.

1.3.1 Distance, Defined in an Inertial Frame

Distance is a very useful notion in many domains. However, we will see that despite its very common usage, its precise definition is not straightforward and relies on axioms. A classical way to define distance is to select a standard rod and then to replicate it identically, so that the distance between any pair of points A and B is obtained by placing such rods end to end along the segment of straight line between A and B, and counting the rods needed. To improve the precision, one should use smaller rods. (For practical reasons, these can be marked as decimals in larger rods, like with rulers.)

We have assumed that A and B are fixed in an inertial frame, and that the observer measuring the distance AB is also fixed in this frame. Distance being so defined, it is possible to build Cartesian orthonormal frames with three perpendicular axes, thanks to the property that a perpendicular line is the set of points on a plane that are at equal distance from two points.

However, this classical method relies on axioms: first, we need to use straight lines, but we know since Euclid that the straight line relies on axioms. Then, we must make sure that two identical rods located in two different points have the same length, and that their lengths remain the same at all times. We will successively address these issues.

1.3.1.1 The Straight Line and the Distance

A classical Euclidean axiom defining the straight line is: "A straight line is the shortest path between two points". However, we notice a circular reference between the notion of distance and the notion of straight line. Arguably, the definition of a straight line should enable us to draw a straight line, and to check if a given line is straight. This axiom doesn't fulfill this requirement because how can we be sure that a shorter path doesn't exist? Actually, this axiom is a characteristic property of a straight line.

Nevertheless, physics gives us a practical way to obtain straight lines, and this is thanks to the fundamental **law of inertia**: any object which is not subject to any external force moves along a straight line and with constant velocity. This law also applies to the photon; hence, laser beams provide us with a practical way of having straight lines.

We can thus see that the notions of distance, straight line and the law of inertia are closely linked together. These notions are primitive, meaning that one cannot define them without invoking other notions which in turn request the initial one, leading to circular references[10].

1.3.1.2 Distance Independence of the Location and the Time

How can we be sure two identical rods measure the same distance independently of their locations? Similarly, how can we be sure the length of a rod remains the same at all times? To answer these questions, we need **the universal homogeneity and isotropy postulates: all properties of the universe are the same everywhere and remain unchanged at all times (homogeneity), and in all directions (isotropy).**

Remark 3: If you are in a region of the universe that incurs an expansion, meaning that all distances are expanded by the same coefficient, there is no way to notice it.

Remark 4: When measuring the length of an object, one must transport standard rods; hence, we need to assume that this transportation will not alter its length. Alternatively and in a more modern way, we can use the same physical phenomenon in different points, such as the wavelength of radiation emitted by an identical atom.

1.3.1.3 Practical Limitations

Historically, precision and stability were sought by several means: using very stable materials such as platinum for building standard rods; then selecting supposedly stable natural distances: the meter was defined as a precise fraction of the Earth meridian.

Notably, the 20th century brought fantastic technological improvements in the precision of the measurements of distance and especially time, so that today the official distance definition uses time as we will see. *Hence, let's address the time definition:*

1.3.2 Time, Defined in an Inertial Frame

We will first consider the time at one point, then in a whole inertial frame.

1.3.2.1 Time in One Point of an Inertial Frame

Time is a more abstract notion than distance: one cannot visualize nor manipulate time. Philosophers have known for a very long time that the notion of time resists precise definition, despite its very common usage.

The word "time" comes from the Latin "tempus" and the Greek "temnein": to cut. Time flows and we want to cut its passage into equal sections that we can count, enabling us to measure the time. This method requires that the sections are perfectly equal, which seems realistic since many periodic phenomena are observed in the universe, such as day/night with for example the exact time when the Sun is in its zenith, but also seasons, years, positions of planets, etc.

However, time cannot be defined without circularly referring to another notion related to time, meaning that time is a primitive notion. For instance, in the above time definition, to be periodic is precisely to happen regularly in time. We are then forced at this step to consider time as a primitive notion and to come up with an axiom: the existence of periodic phenomena having an absolutely stable period. This enables us to define an ideal clock (or perfect clock), which shows the time *at the point where it is located*. Indeed, some variations are a priori possible when the observer is not next to the clock, or is moving relative to the clock as we saw with Einstein's train scenario.

Besides, Time, unlike distance, has an important property: it has a direction. Time distinguishes between the past, the present and the future. Due to this direction, we experience the principle of causality: a consequence cannot precede its cause. This aspect will be further addressed.

Historically, mankind has shown great ingenuity in creating devices to measure time: the sundial, the hourglass, the pendulum, watches, etc. We even can compute the periods of some systems based on known physical laws (for example, the period of a pendulum). This diversity is useful to cross-check the measures of time, and to make sure they don't drift. However, if there is a change in all durations of all phenomena, we wouldn't have a way of noticing it. Hence, as we saw with the distance, we need to postulate the universal homogeneity to make sure that the universe properties regarding the time remain unchanged everywhere and at all times.

To reach extremely high precision, one should measure periodic phenomena that are as brief as possible. In this regard, the 20th century brought incredible enhancements, thanks to scientific breakthroughs in quantum mechanics and technological ones in microelectronics, so that the official time unit, i.e., the second, was officially defined in 1967:

> The second is the duration of 9,192,631,770 periods of the radiation corresponding to the transition between the two hyperfine levels of the ground state of the Cesium-133 atom.

The precision of the best atomic clocks is in the range of 10^{-18} second, and such an atomic clock will be off by up to a second every 10 billion years!

At this stage, we have seen how we can define the time at a given point of an inertial frame. We will now see how we can extend this definition within a whole inertial frame.

1.3.2.2 Time in a Whole Inertial Frame

In order to define time in a whole inertial frame, we must solve the two following issues:

> *How can we be sure that two identical clocks placed in two different locations will beat at the same pace?*

> *How can we be sure both clocks are synchronized? (Meaning that they display exactly the same time.)*

1.3.2.2.1 *How Can We Be Sure That Identical Clocks in Two Different Locations Will Beat at the Same Pace?*

We again need the universal homogeneity postulate, stating that the universe's properties are the same everywhere and at all times. Consequently, identical clocks which are fixed in the same inertial frame must function the same way independently of their locations, meaning that their periods are identical.

1.3.2.2.2 *The Next Question Is: How Can We Synchronize Clocks?*

First, there is a simple way to check if two clocks in two points A and B are synchronized: we place an observer in the middle of the segment AB, and if he sees both clocks continuously displaying the same time, then they are synchronized.

We now want to synchronize two clocks: we can think of building two identical and synchronized clocks in one point A, and then bringing one clock to the point B. However, we will be surprised to discover that, on arriving at B, the two clocks will no longer show the same time. We will indeed see that the acceleration incurred by the second clock during its transportation from A to B will de-synchronize it from its partner.[11]

Fortunately, another method for synchronizing clocks exists: consider any couple of fixed points A and B in an inertial frame K. We place a clock in A equipped with a flash emitter that flashes at every second (Figure 1.3).

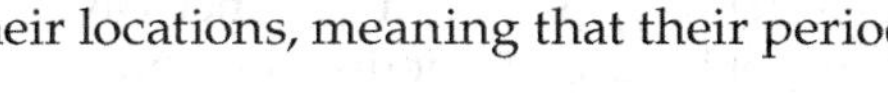

A•))   •B

FIGURE 1.3
Time synchronization by broadcasting a signal.

These flashes reach B with a certain delay T due to the flash propagation at the speed of light. This delay can be calculated: T = AB/c, and it is always the same since A and B are fixed in K.

The flashes periodically arrive at B; hence, they can constitute a clock mechanism in B. Then to synchronize this clock in B with the one at A, we just need to add the delay T to the time corresponding to the number of flashes received at B. It is this way possible to define time within a whole inertial frame. This method relies on the universal homogeneity and isotropy postulates, as well as the inertia law so that we are sure that light propagates along a straight line, and at the same speed in all directions.

1.3.2.3 **The Official Distance Unit Definition Based on the Time**

Time being the physical quantity that can be measured with the greatest precision, we can measure distance with the highest accuracy by using time measurements: to measure the distance between two points A and B, we indeed place two synchronized clocks at these points, and we send a light pulse from A to B. The travel time T of this pulse gives us the distance AB: AB = cT.

Hence in 1983 the international unit of distance, i.e., the meter, was officially defined to be **the distance light would travel (in a vacuum) in precisely 1/299,792,458 seconds.**

Remark 5: This is actually a practical method to achieve very accurate measurements, but the definition of the concept of distance remains as previously described. (Indeed, the speed c is a distance divided by a time; hence,
the notion of speed cannot be used to define the notion of distance.)

*

*One important conclusion is that the classical notions of time and distance remain
valid as long as we consider fixed points within the same inertial frame. We saw that
these notions are based on the universal homogeneity and isotropy postulates, the law
of inertia and some basic axioms defining the straight line and the time at one point.*

*We will now examine the context with different inertial frames in motion relative to
one another, which is the general case in life. We previously saw with Einstein's train
scenario that these notions of time and distance cannot apply unmodified in different
inertial frames. Fortunately, three years after Einstein published his famous papers,
H. Minkowski proposed new concepts regarding time and space which are also valid
in the general context of moving inertial frames, and which proved to be very useful:*

1.4 Minkowski's Powerful Concepts: The Event and the Proper Time

*H. Minkowski famously said: "**Space by itself, and time by itself, are doomed
to fade away into mere shadows, and only a kind of union of the two will
preserve an independent reality**[12]". He then presented two key notions that are
meaningful independently of the frame where they are considered: the **Event** and
the **Proper Time**. These two concepts will be widely used in Relativity (Special and
General); hence, we will present them carefully:*

1.4.1 The Event

We saw with Einstein's train scenario that time and distance are linked
together: particularly, we saw that if we place synchronized clocks at every
window of a moving train, meaning that they display the same time for the
passengers, observers on the platform see all these clocks displaying different times. Consequently, the sole information of the time inside the train
and its speed is not sufficient for an observer on the ground to tell the time
displayed by a given clock inside the train: we need the additional information regarding the location of this clock. This set of information, the time
displayed by a given clock and its location, is called an **event**, and we will

further see that it is sufficient for observers on the platform to calculate the time at the different windows of the train.

More generally, any phenomenon can be characterized in any inertial frame K by its location with three spatial coordinates (x, y, z) and the time t of K at which it has occurred. Minkowski assured this can always be done since "**no one has ever seen an object any other way than at a given time**". If the object is not a single point, it can be characterized by a set of points, and thus a set of events. Hence for the sake of simplicity and clarity, we will further consider the case of point objects.

Time and Distance being linked together, Minkowski stated that we are in a **space-time universe** with 4 dimensions (3 for space and 1 for time), with the event being its elementary item. Before Relativity, the time dimension was independent of the space ones: we were in a (3 + 1)D universe; now, we are in a **4D space-time continuum.**

The trajectory of a point object is fully characterized by a succession of events; such a 4D trajectory is also called the "Worldline" in Relativity.

Mathematically, an event can be considered as a point having four coordinates (x, y, z, t) in a 4D affine space within an inertial frame K. Thanks to the universal homogeneity postulate, the origin-event of a frame can be chosen arbitrarily as well as the directions of its axes. For the sake of simplicity, we will use 4D orthonormal frames, named "Minkowski frames", but this is not an obligation.

An event exists independently of the inertial frame where it is considered: this means that it is an *intrinsic* notion[13]. There is a one-to-one relationship between the coordinates of an event in one inertial frame and those in another inertial frame, and it is the famous Lorentz transformation which we will demonstrate in the next chapter.

Remark 6: When assessing the coordinates of a distant event, one should pay attention that there is a difference between the time at which the observer sees the event, and the time at which this event has occurred: indeed, photons emitted by the considered event need time to reach the eye of the observer. This difference is usually negligible; it is important, however, when considering distant stars or planets, or even with Earth satellites, such as those of the Global Positioning System (GPS).

1.4.1.1 The Space-Time Separation Four-Vector

Consider two events M* and N* which are, respectively, seen in an inertial frame K at the 4D points M and N of coordinates: (Xm, Ym, Zm, Tm) and (Xn, Yn, Zn, Tn). The space-time separation between M* and N*, denoted by M*N*, is seen in K by the vector:

$$\overrightarrow{MN} = [(Xn - Xm), (Yn - Ym), (Zn - Zm), (Tn - Tm)].$$

We will further show (cf. Section 2.2.1) that if two space-time separation vectors are equal in one inertial frame K, meaning $\overrightarrow{MN} = \overrightarrow{PQ}$, then they are equal in all inertial frames (even though in general both will be changed). Such a vector is thus intrinsic and called a "Four-Vector".

Likewise, the origin-event of a frame being arbitrary, if we choose the origin-event of K to be M*, then the coordinates in K of the space-time separation $\overrightarrow{MN}$ are the same as the ones of the event N*.

1.4.1.1.1 Notation Considerations

It is useful to mark differently an event and the point where it is seen in a given frame at a given time. Hence, we will often make this distinction with the symbol *: M* represents the event and M the point. However, other conventions exist.

1.4.2 The Proper Time and the Proper Distance

1.4.2.1 The Proper Time

We saw that the time by itself has no precise meaning for moving observers relative to the clock giving the considered time. Hence, we must attach to the time the indications of where this time is originated and where it is observed (or considered); this is always feasible as "**No time has ever been observed any other way than at a given place**" (Minkowski).

A special place to observe the time of a clock is the location of the clock itself since the time observed there is not affected by any distance or motion relative to the clock. We will use the terms "official clocks" or "perfect clocks" meaning that such a clock gives a second which is identical to one given by the official time definition based the atomic clock as seen previously. Hence, the proper time definition: **the proper time in a point A is the time seen by an observer located in A and watching a perfect clock located in A**. The proper time is usually denoted with the Greek letter τ (pronounced "tau"); subsequently, τ_A means the proper time of a perfect clock located at the point A (and seen by an observer also located in A).

The proper time is an extremely important concept: first, we saw that the very time definition relies on the existence of periodic phenomena seen in one point; then this time in one point could be extended to a whole inertial frame, thanks to a synchronization method. The proper time thus corresponds to the *primitive* notion of time from which the time of an inertial frame is built. Second, the proper time has two fundamental properties: it is the **same in all inertial frames** and it is **invariant**. Let's explain these important properties:

The proper time is the same in all inertial frames: a perfect clock beats at the same pace whether it is on the ground frame or on the train frame, as is seen by an observer co-located with it (or in the same inertial frame). This property stems from the inertial frames equivalence postulate in conjunction with the universal homogeneity and isotropy, as shown in Volume II Section 1.2.

It is then possible in any inertial frame to use the same unit of time, e.g., the second, and this second represents the same time duration for all observers who are fixed in these inertial frames.

The proper time is invariant: in Relativity, **a quantity is said invariant if it has the same value in all inertial frames.** The proper time in a point A, τ_A, is invariant due to the very definition of the proper time. An observer B moving relative to A has no impact on the way the observer fixed in A sees the clock in A. The proper time in A can be used to label events occurring in A at different times, and these labels are identical for observers in any frame.

The proper time has the important compelling interest of being one's biological time: the time perceived by your body wherever you are, whatever your speed and acceleration. Your proper time is the time which specifically determines your aging effects: you may travel in space, the time affecting your body is your proper time, and you can see it with your own watch.

1.4.2.2 The Proper Distance

By analogy with the proper time, the **proper length of a segment is defined as the length of that segment measured by an observer who is fixed relative to this segment.**

Similarly as the proper time, two identical segments, placed in two different inertial frames, have their proper length identical. It is then possible in all inertial frames to use the same unit of distance, e.g., the meter, and this meter represents the same length for all observers who are fixed in these inertial frames.

Consequently, it is possible to compare speeds in different inertial frames (the speed being a distance divided by a time), which gives a meaning to the famous postulate: light propagates at the speed c in all inertial frames, whatever the speed of its source.

The proper length is also **invariant** due to its very definition: an observer B moving relative to an observer A who is fixed in K has no impact in the way A measures a segment that is fixed in K.

Remark 7: When a segment AB is moving relative to an observer fixed in K, the latter measures this segment by considering two events A* and B* that are **simultaneous** in K, with A* occurring at the point A, and B* at the point B. The length in K of this moving segment is then the spatial part of the components in K of the space-time separation Four-Vector A*B*.

Remark 8: The event, the proper time and the proper distance are intrinsic notions as further explained in Volume II Section 1.3.2.

We will now recapitulate the main postulates of Relativity, and elaborate on them so that they will be the basis on which all the laws and results of Relativity will be mathematically demonstrated.

1.5 The Main Postulates of Special Relativity

1.5.1 The Universal Homogeneity and Isotropy

The universal **homogeneity** postulate states that **all universe properties are the same everywhere and remain unchanged at all times.**

The **isotropy** postulate states that **universe properties are the same in all directions in space.** Thus, there is no privileged direction.

Our universe indeed appears to be both homogeneous and isotropic; hence, we will often regroup these two properties under the word "homogeneity".

Mathematically, homogeneity means that the laws of physics are invariant under translation in space and in time; isotropy means that they are invariant under rotation in space.

In addition, the universe is assumed to be a **4D space-time continuum,** with the elementary unit being the event, and where objects trajectories are continuous and even differentiable.

The basic notions of time and distance are based on axioms stating the existence of periodic phenomena in a point (the proper time), and on the Euclidean ones regarding the distance (being based on the straight line).

1.5.2 The Inertial Frames Equivalence Postulate

This postulate, initially called "the principle of Relativity", was presented in Section 1.2.1, and states:

All natural phenomena can be described according to the same physical laws in all inertial frames.

Einstein added that there is no privileged frame, no absolute frame giving, for instance, the true time. He also added that this principle is a postulate, meaning that it doesn't suffer any exception, whereas a principle can.

The inertial frames equivalence postulate implies in conjunction with the homogeneity and isotropy postulates the following important results which will be necessary for demonstrating the Lorentz transformation:

1. The proper time is the same in all inertial frames.
2. The proper distance is the same in all inertial frames.
3. The speed of light is the same in all inertial frames.
4. If K and K' are two inertial frames, the speed of the K relative to K' is the opposite of the speed of K' relative to K.
5. Any physical experiment gives the same results in all inertial frames.

The demonstrations of these fundamental results are given in Volume II Section 1.2, and the following comments show their importance:

- Statement 1 means that two identical clocks in two different inertial frames beat at the same pace, as seen by observers located in front of these clocks.
- Statement 2 is similar to Statement 1. From these two statements, we can define common reference units for the time (e.g., the second) and of distance (e.g., the meter) in all inertial frames.
- Statement 3 stems from the fact that light propagation is a physical phenomenon; hence, a light pulse emitted from a fixed point in one inertial frame obeys the same law as in any inertial frame. However, in non-inertial frames, the speed of light can be greater than c.
- Statement 4, also referred to as the inertial frames velocity reciprocity, is not as straightforward as it may seem. Besides, it is not valid in General Relativity.
- Statement 5 is the essence of the inertial frames equivalence postulate.

Regarding the inertial frame definition, it relies on the law of inertia, which actually is a postulate: **a body that is not submitted to any force continues in a state of rest or in uniform motion along a straight line.**

1.5.3 The Light Speed Invariance Postulate

This postulate issued by Einstein states: **the speed of light is the same in all inertial frames, and independent of the speed of the source which has emitted it.**

We will further see that c is the only invariant speed, and also the maximum speed which can only be reached by massless objects.

**

These are the main postulates on which Special Relativity was built. However, we will need additional postulates when addressing the case of accelerated trajectories and the laws of dynamics. Then when entering into General Relativity, new postulates will be added, and those of Special Relativity will be revised.

1.6 Questions and Problems

The following questions and problems will relate to the following context: you are in a high-speed train moving at constant velocity of 300 km/h relative to the ground, and you have a phone with you so that you can communicate with your friend who is fixed on the platform of a railway station.

Problem 1.1: It is nighttime, you can see nothing by the windows, and hear no noise.

Q1: How can you be sure your train is moving?

Q2: It is now daytime, you pass along the railway platform and you see your friend. Your friend sees you moving at 300 km/h; at which speed do you see your friend moving (backward)?

Q3: Your train is still moving at 300 km/h relative to the ground. You see on the rails next to yours another train passing in the opposite direction than yours. You know that this train is moving at 300 km/h relative to the ground. Is the speed of this train relative to you equal to 600 km/h?

Q4: You have a laser that you point perpendicularly to you and you emit a light pulse.

Q4A: Will your friend see the pulse going perpendicularly to the train?

Q4B: Same question, but imagine yourself at a time before Galileo.

Problem 1.2: When you are exactly in front of your friend, he tells you by phone that there is a tunnel 3,000 m ahead of the train, along a straight line.

Q1: For you in the moving train, is the distance to the tunnel 3,000 m?

Q2: Is the distance from you to the tunnel a proper distance?

Q3: Is the distance from your friend to the tunnel a proper distance?

Problem 1.3: At the exact moment when you are at the level of your friend, you pass your hand through the window, you point your laser toward the tunnel, and you emit a light pulse. At the same moment, your friend also points his laser toward the tunnel and emits a light pulse.

Q1: Your friend measures your pulse speed: Will he find that your pulse goes more rapidly than his?

Q2: Will the two pulses arrive at the tunnel simultaneously?

Q3: Will your pulse take the same time to reach the tunnel as your friend?

Q4: Same questions, but imagine yourself at a time before Einstein.

Problem 1.4: Your friend calculates how much time it will take you to reach the tunnel from the moment you were in front of him.

Q1: What value does he find?

Q2: You mark the time t1 of your watch when you are in front of your friend, and this defines the event E; then you mark the t2 when you reach the tunnel, and this defines the event F. Does (t2 − t1) match with the value calculated by your friend for the time between the events E and F?

Q3: Which is a proper time duration: Is it (t2 − t1) or the value calculated by your friend?

Problem 1.5: You are now hungry and you go to the restaurant car and you just sit. You ask for boiled eggs and the cook asks you how long you want your eggs to be boiled. You usually like them 3 minutes, and you want your eggs to be cooked exactly in the same way, but you wonder if you should ask for the same 3 minutes duration since you are in a fast-moving train and you heard about time Relativity.

Q1: Should you ask for a different duration than 3 minutes?

Q2: Let's assume the answer to the previous question is: X minutes. Is this X minutes a proper time duration?

Q3: You start a phone conversation with your friend when the eggs are put in the boiling water, and you end the call after the X minutes. For your friend, will this conversation last X minutes?

Problem 1.6: Universal homogeneity

If our universe were finite, and with its 3D spatial part being the volume limited by a sphere,

Q1: Could our universe be homogeneous?

Q2: Could it be isotropic?

Notes

1 Quoted by Ronald W. Clark in *Einstein, The Life and Times*, page 73.
2 It is only in the 20th century that we discovered that stars also are moving.
3 Cf. more in Section 8.2.
4 Maxwell's equations are presented in Volume II Section 5-1.
5 Cf. more in Volume II Section 1.2.3.
6 This point will be further developed in Section 4.1.1.3.
7 The stellar aberration phenomenon is presented in Volume II Section 1.1.1.
8 This is further explained in Volume II Section 1.2.1.
9 More explanations are given in Section 8.2.
10 Cf. more in Volume II Section 1.3.3.
11 The famous twin paradox is built on this phenomenon, which we will see in Section 3.5.

12 Pronounced by H. Minkowski at the beginning of his talk at the 80th Assembly of German Natural Scientists and Physicians, on September 21, 1908. Quoted by Ronald W. Clark, *Einstein, the Life and Times*, page 123.

13 Cf. more in Volume II Section 1.3.2.

Quoted in the Book "Einstein et la Relativité. *L'espace est une question de temps*", by David Blanco Laserna, French edition, page 94.

2

The Lorentz Transformation

Introduction

The Lorentz transformation will be demonstrated as a consequence of the revolutionary time dilatation law. The latter will be demonstrated using a famous scenario imagined by Einstein, but before that, an important prerequisite will be shown: the invariance of the transverse distance. The length contraction law will then be deduced. The linearity of the desired 4D function, which transforms the four coordinates of any event from one inertial frame to another, will be shown. Consequently, the number of parameters of this new function will be reduced from 16 to 4. This new function, i.e., the Lorentz transformation, will then be demonstrated using two simple scenarios involving the physical laws previously shown. We will then see direct applications and a very useful tool to visualize scenarios and solve problems: Minkowski diagrams.

2.1 Time Dilatation and Other Fundamental Laws

2.1.1 Distance Invariance along the Transverse Directions (y, z)

Einstein's train scenario showed that time and distance are relative to a frame. However, looking more precisely at this scenario, the Relativity of simultaneity concerned points that are parallel to the motion of the train (the points A and B on the platform); hence, we cannot rigorously deduce that Relativity also concerns segments that are perpendicular to the train motion. The following scenario will show that these transverse distances are invariant, meaning equal in all inertial frames.

2.1.1.1 The Tennis Ball Scenario Showing the Invariance of Transverse Distances

Picture a tennis ball and a solid wall having a hole of exactly the size of the ball, this size being measured when the ball is fixed relative to the wall. (We can place the ball in the hole and see that it perfectly fits in the hole.) We then go backward and throw the ball toward the hole extremely fast, and in a perpendicular direction to the wall, targeting the center of the hole. We assume the ball moves along a straight line toward the center of the hole. We will neglect the effects due to gravity and to the acceleration phase (Figure 2.1).

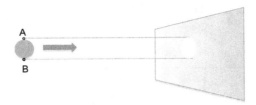

FIGURE 2.1
The tennis ball going toward the hole perpendicularly to the wall.

The segment AB is a diameter of the ball in a perpendicular direction to the ball motion. Let's make a hypothesis (h1) that the Relativity of distance also concerns segments that are perpendicular to the motion of one frame relative to the other. We then place ourselves in the frame of the wall, and this hypothesis (h1) implies that the segment AB has a different size than when the ball was fixed, meaning different than the hole size. Hence, we wonder whether the ball is smaller, in which case it will pass through the hole, or larger and the ball won't traverse the wall.

Let's then take a second hypothesis (h1-a): an observer sees a moving object (relative to him) smaller than when it was at rest. Consequently, our observer in the wall frame will see the moving ball smaller than when it was fixed; hence, the ball will traverse the wall.

Let's now have a second observer who is fixed relative to the ball frame. He sees the hole coming toward him, and according to the hypotheses (h1) and (h1-a), the hole size is smaller than when the wall was fixed. He will then conclude that the ball won't traverse the wall.

These two statements are contradictory and incompatible because there is only one reality that can be checked experimentally: either the ball traverses the wall or not. One of the two observers must then be wrong, meaning that at least one assumption that he used was wrong.

Let's assume that it is the hypothesis (h1-a) which was wrong: it implies that the following alternative assumption denoted by (h1-b) is true: "An observer sees a moving object larger than when this object was fixed relative to him". However, this assumption (h1-b) leads to a similar problem: the first observer fixed with the wall will say the ball is larger and so it won't traverse the wall, whereas the second observer in the ball frame will say the hole is larger and so the ball will go through. Consequently, it is the first hypothesis (h1) which must be wrong, meaning that distances are invariant along the transverse directions.

Remark 1: The fact that the observer on the wall frame and the one on the ball frame make contradictory statements is not a problem by itself: Einstein's train scenario gave us an example where observers in different frames are entitled to issue contradictory statements about simultaneity. The difference is explained in Volume II Section 1.3.1, showing that the contradiction is apparent in this scenario, whereas it is real in Einstein's train scenario.

Remark 2: We made some implicit assumptions in this scenario such as: A) a straight line in K is also seen as a straight line in K'. B) A transverse direction in K is also seen as a transverse direction in K'. The reader may admit at first these results which seem natural, and then see their demonstrations: the answer to A) is given with Problem 2–5. The answer to B) is given in a more general demonstration of the invariance of transverse distance presented in Volume II Section 1.4.2. This latter demonstration also shows the non-relativity of time along the transverse directions.

2.1.2 The Fundamental Time Dilatation Law

We will use the following simple and luminous scenario imagined by Einstein to demonstrate this revolutionary law. In a first step, we will build a perfect clock using a light pulse traveling between two mirrors. In a second step, we will place this clock mechanism in a frame K' moving relative to the frame K of an observer, and the two mirrors will be placed along a transverse direction so as to benefit from the invariance of transverse distances. This observer will calculate the duration taken by the pulse to make the round trip between the two mirrors, and this will give us the famous time dilatation law.

2.1.2.1 Einstein's Perfect Clock

In an inertial frame K', we place two mirrors face to face at both ends of a segment O'M' along the O'Y' axis. We then emit a light pulse from O' toward M': it will reach M' where it will be instantaneously reflected and will return to O'. The mirror in O' will instantaneously reflect this pulse toward M', and this scenario will repeat itself again and again, providing a periodic phenomenon which we will use as a perfect clock (Figure 2.2).

FIGURE 2.2
Einstein's perfect clock.

The point O' periodically receives the light pulse, and this constitutes a perfect clock in O': by counting the number of arrivals of the pulse, this clock displays the time from the first pulse emission. The proper period of this clock, denoted by $\Delta\tau_{0'}$, is the duration between two consecutive arrivals of the pulse at the point O'. This proper period can be easily calculated:

$$\Delta\tau_{0'} = 2OM'/c. \tag{2.1}$$

If we use the international system of units for the distance O'M' and the speed c, then $\Delta\tau_{0'}$ is expressed in seconds. Besides, we place in O' a counter of the number of pulses received, and at any moment, the number displayed by this counter enables us to know the time in seconds elapsed from the first pulse emission till that moment: if it displays the number N, the time elapsed at this moment is: $N\Delta\tau_{0'}$.

2.1.2.2 The Famous Time Dilatation Scenario

Let's now place ourselves in an inertial frame K where the previous frame K' hosting the perfect clock is seen moving at the speed v along the direction of the OX axis of K. The segment O'M' is constantly along a perpendicular direction relative to the OX axis of K and O'X' of K'. We set that both frames K and K' share the same origin-event O* which is: "The point O' of K' coincides with O of K". Besides, at this event O*, the point O' of K' emits a light pulse along O'Y', and this pulse will reach the mirror in K' at the point M'.

Consider an observer fixed in K; he first sees the pulse covering the segment ON, where N is the fixed point of K which coincides with M' when the pulse reaches M'. He then sees the pulse covering the segment NB, where B is the fixed point of K that coincides with O' when the pulse reaches O' after having been reflected in N.

This observer can calculate (and also measure) the total duration for the pulse to cover this trip, and this duration will give him the period of the perfect clock, as seen by anyone in the frame K. Then, by comparing this period, denoted by ΔT, with the proper period of this clock, $\Delta\tau_{0'}$, we will obtain the desired time comparison law.

Remark: The mirror in O' is not represented in Figure 2.3 for the sake of clarity.

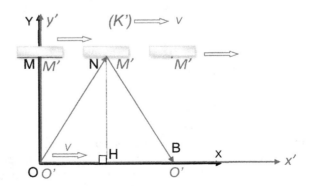

FIGURE 2.3
The perfect clock is moving with respect to an observer in O.

Calculation of ΔT: Thanks to the light speed invariance postulate, our observer in O sees the pulse going at the speed c. We thus have:

$$\Delta T = ON/c + NB/c.$$

Conveniently, taking into account the symmetry of the figure relative to HN, the point H being the orthogonal projection of N on the OX axis, we have:

$$\Delta T = 2ON/c. \tag{2.2}$$

Let's now apply the Pythagorean relation in the triangle OHN:

$$\mathbf{ON^2 = OH^2 + HN^2}. \tag{2.3}$$

We have OH = ½ OB, where OB is the distance covered by the point O' during the duration ΔT, and we know that the speed of O' is v; thus:

$$OH = \tfrac{1}{2}\Delta Tv. \tag{2.4}$$

Likewise, HN = OM, where M is the fixed point of K which coincides with M' at the time t = t'= 0. Then, thanks to the invariance of transverse directions, we have: OM = O'M', where O'M' has the same length as in Section 2.1.2.2 when this segment was considered fixed. Hence with equation 2.1, we have:

$$HN = \tfrac{1}{2}\Delta\tau_{0'}\, c. \tag{2.5}$$

Note that we used the fact that the proper period $\Delta\tau_{0'}$ has a precise meaning in any frame, being intrinsic and invariant (cf. Section 1.4.2).

Let's return to equation 2.3, and from equations 2.2, 2.4 and 2.5, we obtain:
$(\tfrac{1}{2}\Delta T\, c)^2 = (\tfrac{1}{2}\,\Delta T\, v)^2 + (\tfrac{1}{2}\,\Delta\tau_{0'}\, c)^2$.

We can then express ΔT: $\Delta T^2\,(c^2 - v^2) = \Delta\tau_{0'}\, c^2$, which finally gives:

$$\Delta T/\Delta\tau_{0'} = \sqrt[2]{\frac{1}{1 - v^2/c^2}}\,.$$

This fraction is classically denoted by the Greek letter "gamma":

$$\gamma_v = \sqrt[2]{\frac{1}{1 - v^2/c^2}}\,.$$

So the result finally is: $\Delta T = \gamma_v\,\Delta\tau_{0'}$ ■

The table which follows gives the values of γ_v for different values of the speed v. We first notice that this gamma factor is always greater than 1, hence the time dilatation law:

The proper period of a clock which is moving relative to an inertial frame K at the velocity v is seen dilated in this frame K by the gamma factor γ_v.

We notice that the gamma factor is extremely close to 1 for usual speeds, as shown in the table below. It is only for speeds higher than 15 million km/h that the difference between γ_v and 1 is greater than 0.01%. This explains why we don't usually notice the relativistic effects on the time. However, we also notice that when the speed v gets close to c, gamma tends to the infinite: time gets infinitely dilated (Figure 2.4).

Speed km/h	Speed km/s	Speed v/c = β	γ
36 000	10	0.0000334	1.0000000001
108 000	30	0.0001001	1.000000005
360 000	100	0.0003336	1.000000056
3 600 000	1 000	0.0033356	1.000005563
15 000 000	4 167	0.0138985	1.0000966
50 000 000	13 889	0.0463284	1.0010749
155 000 000	43 056	0.1436181	1.0104754
460 000 000	127 778	0.4262215	1.1054
809 438 319	224 844	0.7500000	1.5119
935 000 000	259 722	0.8663415	2.0022
1 017 550 000	282 653	0.9428297	3.0005
1 071 630 000	297 675	0.9929385	8.43
1 076 400 000	299 000	0.9973583	13.77
1 079 100 000	299 750	0.9998600	59.76
1 079 200 000	299 778	0.9999527	102.77
1 079 251 000	299 792	0.9999999	2421.88
1 079 251 092	299 792	1.0000000	∞

FIGURE 2.4
The gamma factor table.

For usual speeds, v/c is very small so that γ_v can be approximated by the first terms of its series expansion (Taylor-Young), which are: $\gamma_v \sim 1 + \frac{1}{2}(v/c)^2$ (cf. more in Volume II Section 1.5.2). The ratio v/c is denoted by β, and called the "Lorentz Factor"; it is frequently used in Relativity

2.1.2.3 Generalization and Consequences

After having reached the point B of K, the pulse will be reflected and directed toward the mirror in M′ and the scenario will repeat itself. The second round trip leads to the same conclusion as the first one: if we place an observer fixed in B of K, he will see an identical scenario as the first one which is depicted by Figure 2.3, so that the proper period of the clock in O′ is always seen dilated in K by the same gamma factor γ_v. (We assumed K′ always moves at the velocity v relative to K.)

This enables us to tell how the proper time in O′ is seen from the frame K. We again assume that K and K′ share the same event-origin O* (as previously defined). Let's then consider an event F* which takes place in O′ at exactly the Nth arrival of the light pulse. The perfect clock in O′ displays the value N, meaning that the proper time in O′, denoted by $\tau_{0'}(F*)$ is: $\tau_{0'}(F*) = N\Delta\tau_{0'}$.

In K, this event F* is also seen at the Nth arrival of the light pulse at O′, meaning at the time denoted by T(F*) and which is: $T(F^*) = N\gamma_v \Delta\tau_{0'} = \gamma_v \tau_{0'}(F^*)$. This shows that **any proper duration at one point which is moving relative to an inertial frame K is seen in K dilated by the gamma factor.**

Another important property is derived:

2.1.2.3.1 The Chronological Order between Two Co-Located Events Is Invariant

Consider two events F* and G* which both take place at the point O′ of K′; in this frame K′, the event F* is seen occurring before G*. The question is: in another inertial frame K, is the event F* also seen occurring before the event G*? The answer is positive because we have from the previous relation: $T(F^*) = \gamma_v \tau_{0'}(F^*)$ and $T(G^*) = \gamma_v \tau_{0'}(G^*)$. Then, as $\tau_{0'}(G^*) > \tau_{0'}(F^*)$, we have: $T(G^*) > T(F^*)$.

Remark: This property of **chronological order invariance between two co-located events** may seem natural, but we will further see that it is not always respected if the two events are not co-located. It is essential that it is respected in the case of co-located events because the first event F* can be the *cause* of the second event G*, and it would then be absurd if from another inertial frame, the event G* would be seen occurring before F*.

2.1.2.4 The Time Dilatation Law Is Independent of the Direction of the Clock Motion

One may wonder if the time dilatation law is still valid if, seen from the inertial frame K, the perfect clock is moving in another direction than OX, but still with a constant velocity $\vec{v}$. Besides, at the time t = 0 in K, the point O′ of K′ may not coincide with the point O, but be at a random point A of K (Figure 2.5).

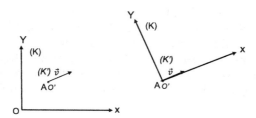

FIGURE 2.5
Time dilatation in case the clock motion is not along OX.

We can change the origin and the axis of K so that its new origin is A and its new OX axis is parallel to the vector $\vec{v}$. We can then place an observer fixed in A of K, so that our previous reasoning applies, leading to the same result: $\Delta T = \gamma_v \Delta\tau_{0'}$ with T being the time of the inertial frame K. The time T of K has not been altered by the change of the spatial axis and the location of its origin.

2.1.2.5 *Beware of Abusive Usage of the Time Dilatation Law*

The time dilatation law cannot be used to transform the time t' of the frame K' into the time t of the frame K independently of the location of the events determining the considered time. In particular, one cannot write $t = \gamma_v t'$ as shown by the following reasoning by absurdity: if this relation were correct, then the principle of Relativity would enable us to write: $t' = \gamma_{-v} t$. (The velocity of K relative to K' is indeed the opposite as the one of K' relative to K.) Then, as the gamma formula involves v^2, we have: $\gamma_v = \gamma_{-v}$, and so: $t' = \gamma_v t$, with the absurd consequence that: $t = \gamma_v^2 t$.

This again shows that the time t of a frame K does not have a meaning in any other frame than K, unless we specify the exact fixed point in K which carries this time, hence the importance of the intrinsic notions of **proper time** and **event**.

Remark: Another demonstration showing the impossibility to write $t = \gamma_v t'$ is presented in Problem 2.4.

2.1.3 **The Length Contraction Law**

We will show that the time dilatation law implies the distance contraction law, and for this, we will use the same scenario as for the time dilatation (cf. Figure 2.3):

Consider the segment OB: it is fixed in the frame K; hence, its length measured or calculated in K is the *proper* length of this segment, which will be denoted by $L_p(OB)$.

Observers in K can calculate the length of this segment OB by considering that the point O' of K' has traveled from O to B at the constant velocity v during ΔT. Hence, they deduce:

$$L_p(OB) = v \, \Delta T.$$

Let's now assess the length of OB seen from the frame K': an observer fixed in O' of K' sees the segment OB going to the left at the speed −v. He first sees the point O at $t' = 0$, and then B at $t' = \Delta \tau_{O'}$. He deduces that O has covered the length OB during $\Delta \tau_{O'}$, and he concludes that the length $OB = v \Delta \tau_{O'}$. This length is not a proper length since the segment OB is moving relative to this observer in O'; hence, we will denote this length by $L_{K'}(OB)$, and we thus have:

$$L_{K'}(OB) = v \Delta \tau_{O'}.$$

Let's apply the time dilatation law: $\Delta T = \gamma_v \Delta \tau_{O'}$. Consequently: $L_p(OB) = \gamma_v L_{K'}(OB)$.

So finally: $L_{K'}(OB) = \dfrac{L_p(OB)}{\gamma_v}$. ∎

This result expresses the length contraction law: **an observer who is moving relative to a segment sees this segment smaller than its proper length, the reduction ratio being the gamma factor.**

Remark: The wording "length contraction" may be misleading: the segment indeed does not shrink; its matter is not compressed; it is the way it is seen by a moving observer relative to this segment, which is different from when this segment was fixed relative to him. This difference is real as it can be measured. A similar situation exists every day when you look at a distant object: you see it smaller than when this object is closer to you. The fact that we do not notice this effect stems only from the fact that we never see objects moving at speeds close to c. The same consideration applies to the time: the "time dilatation" effect does not mean that the time of the considered clock changes; its proper time remains identical, but it is the way it is seen (or considered) by a moving observer which is different.

Remark: Another demonstration is given with Problem 2.16.

2.2 Determination of the Lorentz Transformation

We will always consider the case of two inertial frames, K and K', with K' moving at the constant speed v along the OX and O'X' axes. In a first part, we will assume that both frames share a common origin-event O*. The case of different origin-events will then be addressed in Section 2.2.5.

The Lorentz transformation is the 4D function, denoted by Φ, that gives the 4 coordinates of an event in K' knowing those in K. In order to find Φ, we will first show that this function must be linear. This will enable us to show that it has only four parameters, and these will be obtained by applying the physical laws already found to relatively simple scenarios.

2.2.1 The Lorentz Transformation Must Be Linear

Let's first recall the general definition for linearity for a 4D vectorial function: the function Φ is linear if for any couple of 4D vectors $\vec{u}$ and $\vec{v}$, and for any couple of numbers, x and y, we have:

$$\Phi\left(x.\vec{u} + y.\vec{v}\right) = x.\Phi(\vec{u}) + y.\Phi(\vec{v}).$$

Consider four events A*, B*, C* and D* which are seen in the frame K at the points A, B, C and D. We make the hypothesis that these points are such that: $\overrightarrow{AB} = \overrightarrow{CD}$.

In the frame K', these four events are seen at the points A', B', C' and D'. We will demonstrate that in K' we also have: $\overrightarrow{A'B'} = \overrightarrow{C'D'}$. This property will then enable us to demonstrate the linearity of the function Φ, which will be done in the mathematical complement Section 2.5.1.

The origin-event of any frame being arbitrary, we choose for the sake of simplicity A* as the common origin-event of K and K', as shown in Figure 2.6 which represents these points in 2 dimensions, knowing that the same reasoning applies to 4 dimensions.

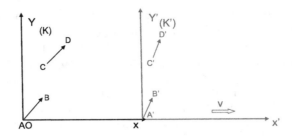

FIGURE 2.6
Two equal space-time separation vectors in K and their images in K'.

We thus have: $\Phi(A) = (A') = (0, 0, 0, 0)$. Then, $\Phi(B) = (B')$. We have denoted by (B') the set of four coordinates of the point B' of K'. Besides the coordinates of the point B are identical to those of the vector $\overrightarrow{AB}$; similarly, those of B' in K' are identical to those of $\overrightarrow{A'B'}$.

The relation $\Phi(B) = (B')$ is thus equivalent to: $\Phi(\overrightarrow{AB}) = \overrightarrow{A'B'}$. Then, having by hypothesis in K: $\overrightarrow{AB} = \overrightarrow{CD}$, we have: $\Phi(\overrightarrow{CD}) = \overrightarrow{A'B'}$.

Let's now choose the event C* to be the common origin-event of K and K', while keeping the CX and C'X' axes parallel to the speed $\vec{V}$. The new frames K and K' are denoted by K^c and K'^c as shown in Figure 2.7:

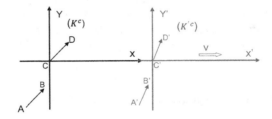

FIGURE 2.7
The frames K and K' are translated by $\overrightarrow{AC}$ and $\overrightarrow{A'C'}$, respectively.

The function which gives the four coordinates of an event in K'^c knowing its coordinates in K^c is denoted by Φ^c, and we a priori don't know if it is the same function as Φ. We, respectively, denote by (D^c) and (D'^c) the four coordinates of the event D* in the frames K^c and K'^c. We thus have: $(D'^c) = \Phi^c(D^c)$.

The function Φ^c must be the same as the function Φ because the universal homogeneity postulate states that all physical laws are the same whatever the location. In our case, the only difference between Φ^c and Φ is the location where the physical law giving the transformation of an event is applied: indeed, the directions of the axes and the velocity $\vec{V}$ of K' relative to K are identical in both cases.

Hence, we must have: $\Phi^c(D^c) = \Phi(B)$. We saw that $\Phi^c(D^c) = \overrightarrow{C'D'}$, and that $\Phi(B) = \overrightarrow{A'B'}$; hence: $\overrightarrow{C'D'} = \overrightarrow{A'B'}$ ∎

We have thus shown that if two space-time separation vectors are equal in one inertial frame K, $\overrightarrow{AB} = \overrightarrow{CD}$, then they are equal in all inertial frames. Such a separation vector is thus intrinsic and called a "Four-Vector". We have also shown that the function Φ which transforms the coordinates of an event is the same as the one which transforms the coordinates of a Four-Vector, provided that K and K' have the same origin-event.

Remark: Another demonstration, quicker but using more complex mathematical means, is given in Volume II Section 1.4.1.

2.2.2 The Function Φ Has Only Four Parameters

An event A* is seen in K at the point A(x, y, z, t), and in K': A'(x', y', z', t'). Any function Φ giving (x', y', z', t') from (x, y, z, t) has the following general form:
    x' = F(x, y, z, t) ; y' = G(x, y, z, t) ; z' = H(x, y, z, t) and t' = K(x, y, z, t).

We just saw that function Φ is linear; consequently, all these 4 functions F, G, H and K must be linear. This means that the function F has the following form: $x' = a_{11}.x + a_{12}.y + a_{13}.z + a_{14}.t$, with a_{1i} being 4 parameters which are only a function of the velocity v of K' relative to K.

The same reasoning applies to the three other functions; hence, there are 16 parameters totally defining the function Φ. For instance, the function K is:
$$t' = a_{41}.x + a_{42}.y + a_{43}.z + a_{44}.t.$$

However, we will see that thanks to the invariance of the transverse distances, the total number of parameters will be reduced to 4:

For the sake of clarity and simplicity, we will make a reasoning with two dimensions, the axis OX and the transverse direction OY, knowing that it will apply to the other directions (OZ and then the time axis). In K, the projection of the point A on the OX axis is the point H. The image of H by Φ is the point H' in K'. We will show that the coordinates of A' in K' satisfy: $y'_A = y_A = HA$, and that: $x'_{A'} = F(\overrightarrow{OA}) = F(\overrightarrow{OH})$ (Figure 2.8).

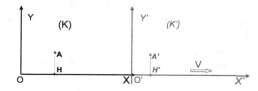

FIGURE 2.8
The event A* seen in K and K', with its projections on OX and O'X'.

First part: The point H being on the OX axis, its image H' is on the O'X' axis. Likewise, the segment HA being along a transverse direction in K, its image H'A' is also along a transverse direction in K'.[1] Moreover, the transverse distances being invariant, we have: H'A' = HA, meaning that: $y'_A = y_A$.

Second part: We have: $\overrightarrow{OA} = \overrightarrow{OH} + \overrightarrow{HA}$; hence: $\Phi(\overrightarrow{OA}) = \Phi(\overrightarrow{OH}) + \Phi(\overrightarrow{HA})$ due to the linearity of Φ.

We then have: $x'_A = F(\overrightarrow{OA}) = F(\overrightarrow{OH}) + F(\overrightarrow{HA})$. We previously saw that the image of $\overrightarrow{HA}$ by Φ is $\overrightarrow{H'A'}$, and that $\overrightarrow{H'A'}$ is along a transverse direction; hence, $F(\overrightarrow{HA}) = 0$. Consequently: $x'_A = F(\overrightarrow{OH})$, which means that:

$$x'_A = F(\overrightarrow{OH}) = a_{11}.x_A + a_{12}.0 + a_{13}.0 + a_{14}.t_A = a_{11}.x_A + a_{14}.t_A.$$

For the sake of simplicity, the parameters a_{11} and a_{14} will be respectively denoted by a and b. The Function F is then: $x' = a_{11}.x + a_{14}.t = a.x + b.t$.

Regarding the function K, the same reasoning applies if we replace the OX dimension with the OT dimension. Likewise, the parameters a_{41} and a_{44} will be replaced by the e and f. (We preferred the letters e and f rather than c and d in order to avoid confusion with the light speed c, and the differential symbol d). Hence, we obtain: $t' = a_{41}.x + a_{44}.t = e.x + f.t$.

Conclusion: The function Φ is such that: $y' = y$; $z' = z$ and:

$$x' = a.x + b.t \tag{2.6}$$

$$t' = e.x + f.t \tag{2.7} \blacksquare$$

In order to find these four parameters, we need four equations. These will be obtained by two scenarios which involve the physical laws previously found, knowing that each scenario will give two equations because we will reason in terms of events.

2.2.3 The Two Scenarios to Obtain Four Equations Involving the Four Parameters

2.2.3.1 Scenario 1

We place a fixed clock at the origin-point O of the frame K. At the time $t = 0$ in K, this clock emits a flash, which constitutes the origin-event for both frames. Then at the time $t = 1$ second in K, the clock emits a second flash, which constitutes the event E*, with its coordinates in K being $(0, 1)$.

Remark: We use the notation whereby the first value corresponds to the abscissa in meter along the OX axis and the second one corresponds to the time coordinate expressed in second.

Let's find the coordinates of E* in K': we notice that the duration in K between the two flashes is a proper time duration in O; hence, in the frame K', this duration is seen dilated by the γ_v factor. Consequently, the event E* is seen in K' at the time: $t' = \gamma_v.1$ seconds.

Seen from K' at that time $t' = \gamma_v$, the point O will have covered the distance: $-v.\gamma_v$ since the velocity of O in K' is: $-v$. Thus, the coordinates of the event E* in K' are: $(-v.\gamma_v, \gamma_v)$.

Let's apply these results to the relations 2.6 and 2.7:

The relation (2.6), $x' = a.x + b.t$, gives: $-v.\gamma_v = a.0 + b.1$; hence:

$$b = -v.\gamma_v. \tag{2.8}$$

The relation (2.7), $t' = e.x + f.t$, gives: $\gamma_v = e.0 + f.1$; hence:

$$f = \gamma_v. \tag{2.9}$$

2.2.3.2 Scenario 2

We place a fixed clock at the origin point O′ of the frame K′. At the time $t' = 1$ second in K′, this clock emits a flash, which constitutes the event F*. The coordinates of F* in K′ thus are: (0, 1).

Let's calculate the coordinates of F* in K: we notice that this 1 second duration in K′ is a proper time duration of the clock in O′; consequently, this proper time duration of 1 second in K′ is seen dilated in the frame K by the γ_v factor; hence, the event F* is seen in K at the time $t = \gamma_v.1$ seconds.

Then in K, at that time $t = \gamma_v$, the point O′ will have covered the distance $v.\gamma_v$. Consequently, the coordinates of the event F* in K are: $(v\,\gamma_v, \gamma_v)$.

We again apply these results to the relations 2.6 and 2.7 and we find:
The relation 2.6 gives: $0 = a\,v\,\gamma_v + b\,\gamma_v$; then, with the relation 2.8:
$0 = a\,v\,\gamma_v - v\,\gamma_v^2$; hence finally: $a = \gamma_v$. $\tag{2.10}$

The relation 2.7 gives: $1 = e\,v\,\gamma_v + f\,\gamma_v = e\,v\,\gamma_v + \gamma_v^2$.

Then: $e.v = 1/\gamma_v - \gamma_v = \gamma_v\,(1/\gamma_v^2 - 1) = \gamma_v\,(1 - v^2/c^2 - 1) = -\gamma_v v^2/c^2$; so finally:

$$e = -\gamma_v v/c^2. \tag{2.11}$$

Remark: Both scenarios involve the time dilatation law; hence, the reader may wonder if they are really independent. Actually, the first scenario also involves another law: the velocity of O relative to K′ is the opposite of the velocity of O′ relative to K (cf. note 2).

2.2.3.3 Conclusion: The Lorentz Transformation

Finally, applying relations 2.8–2.11 to the relations 2.6 and 2.7, we obtain the Lorentz transformation:

$x' = \gamma_v.x - v.\,\gamma_v.t$; then: $x' = \gamma_v\,(x - v.t)$
$t' = -\gamma_v\,v/c^2.x + \gamma_v\,t$; then: $t' = \gamma_v\,(t - x.v/c^2)$ ∎

which constitutes the Lorentz transformation.

*

Note that for usual speeds v/c is very small; consequently, the Lorentz transformation gives extremely close results to the classical Galilean law, which explains why we didn't imagine the latter was wrong.

From now on, we will denote by Λ the function Φ (to remind of Lorentz), or even Λ_v with reference to the speed v of the frame K′ relative to K.

Remark: Given the importance of the Lorentz transformation, three other demon-
strations are presented revealing different aspects. One is based on the symmetry of
the Lorentzian norm (cf. Volume II Section 2.2); one uses the Minkowski space with
complex numbers (cf. Volume II Section 2.4.1); and one shows that the Lorentz trans-
formation is a hyperbolic rotation (cf. Volume II Section 2.2.2).

2.2.3.4 The Reverse Lorentz Transformation

The reverse transformation giving (x, t) knowing (x′, t′) is also a Lorentz
transformation since all physical laws are the same in all inertial frames.
Moreover, we saw that the speed of O of K considered in K′ is −v; hence,
we deduce that the reverse Lorentz transformation is simply obtained by
replacing v with −v in the above Lorentz transformation, which means
that:

$$\Lambda_v^{-1} = \Lambda_{-v}\,.$$

Additionally, we have: $\gamma_{-v} = \gamma_v$ since the gamma factor involves v^2. The reverse
(or inverse) Lorentz transformation is then:

$$x = \gamma_v(x' + v.t') \quad \text{and} \quad t = \gamma_v(t' + x'.v/c^2).$$

2.2.4 The Lorentz Transformation with the Matrix Format

Being a linear function, the Lorentz transformation can be expressed with
the matrix form, which considerably facilitates calculations; we indeed have:

$$\begin{pmatrix} x' \\ t' \end{pmatrix} = \gamma \begin{pmatrix} 1 & -v \\ -v/c^2 & 1 \end{pmatrix} \begin{pmatrix} x \\ t \end{pmatrix}.$$

We will denote by (Λ_v) this Lorentz matrix.
  The reverse Lorentz matrix is simply obtained by replacing v with −v.
The reverse Lorentz matrix, denoted (Λ_v^{-1}), is then:

$$\begin{pmatrix} x \\ t \end{pmatrix} = \gamma \begin{pmatrix} 1 & v \\ v/c^2 & 1 \end{pmatrix} \begin{pmatrix} x' \\ t' \end{pmatrix}.$$

2.2.5 If the Two Frames have Different Origin-Events, the Event Transformation Is an Affine Function

We remind that the Lorentz transformation concerns two different inertial frames, K
and K′, sharing a common origin-event, and having their spatial axes parallel, with

OX and O'X' parallel to the velocity $\vec{v}$ of K' relative to K. In addition, their time and distance units are the same.

We will now consider the case where the two frames have the above characteristics except that they don't share a common origin-event. We will show that the transformation which gives the four coordinates of an event in the second frame, knowing its coordinates in the first one, is an affine function.

Actually, we have encountered this context in Section 2.2.1; hence, we will use the same scenario. The reader is then invited to look at it again, especially Figures 2.6 and 2.7.

We initially have two frames K and K' sharing a common origin-event, A*, and with their axis parallel. We then make a translation in K' so that its origin-event becomes the event C*. Subsequently, this frame is denoted by K'^{c}. Both K' and K'^{c} are moving at the same velocity v̄ relative to K, and their axes are parallel.

Consider a random event D*, with its coordinates, respectively, denoted by (D) in K, (D') in K', and (D^c) in K'^{c}. We want to determine the function Ψ which gives (D'^{c}) knowing (D), meaning: (D'^{c}) = Ψ(D).

In K'^{c}, we have: (D'^{c}) = $\overline{C'D'}$. The coordinates of the vector $\overline{C'D'}$ are the same in K' and in K'^{c} (since K'^{c} only differs from K' by a translation).

Let's introduce the event A*: we have in both K' and K'^{c}: $\overline{C'D'} = \overline{C'A'} + \overline{A'D'}$.

- The term $\overline{C'A'}$ represents the coordinates of A* in K'^{c} : $\overline{C'A'} = (A'^c)$.
- The term $\overline{A'D'}$ is equal to $\Lambda_v(D)$, where Λ_v is the Lorentz transformation from K to K'.

Hence finally: $\Psi(D) = \overline{C'D'} = \overline{C'A'} + \overline{A'D'} = \Lambda_v(D) + (A'^c)$ ∎

This shows that Ψ is an affine function: Ψ is indeed the Lorentz transformation (which is linear) plus a fixed 4D term which is the coordinates in K'^{c} of the origin-event of K.

*

More generally, if the spatial axes of K' are not parallel to the ones of K, then one should proceed in two steps: a first Lorentz transformation with the spatial axes of both frames being parallel, and then a spatial rotation within K' as in classical physics.

Even more generally, one should distinguish between a frame and a system of coordinates. An inertial frame is linked to an inertial object in a rigid manner. All the points of this frame move at the same velocity as this inertial object. Within a given frame, there can be different systems of coordinates, some orthonormal, some not, and some even curvilinear. The directions of the three spatial axes and the origin-event are arbitrary.

All transformations that leave physical laws invariant form the group of Poincaré. These are composed of the Lorentz transformations, the space-time translations and the spatial rotations.

2.3 Validation of the Lorentz Transformation and First Applications

To be valid, any physical law or theory must (i) be confirmed by experiments and observations, and (ii) be consistent with all other laws. We will first see the experimental aspects, and then the theoretical ones.

2.3.1 Experimental Aspects

Unfortunately, at the time when Relativity was found, it was impossible to test the theory by direct experiments involving extremely high speeds. There was, however, one strange phenomenon involving very high speeds which could be explained by Relativity: Walter Kaufmann showed that the trajectories of electrons accelerated at high speeds – as high as within a range of 200,000 km/h – inside cathode ray tubes didn't follow the classical laws. It seemed as if the electron masses had increased, which is normal according to Relativity: the electron inertia increases with its speed (as we will further see). However, other explanations were provided and Einstein did not convince his peers.

In addition, another experiment agreed well with Relativity and consolidated Einstein's confidence in his new theory: the Fizeau experiment, where a light beam passes through moving water[3]. The speed of light relative to the ground conformed with the Relativistic speed composition law. However, many scientists still believed in the ether existence; hence, the relativistic explanation did not convince the majority.

The first important experimental confirmation occurred in 1919: it was the observation of the bending of light in the neighborhood of the Sun, confirming both General Relativity and Special Relativity (the former being based on the latter).

It is much later, in 1932, that the confirmation of the famous law $E = mc^2$ occurred when Cockcroft and Walton succeeded in performing the fission of lithium-7 (cf. more in Section 5.2).

Later still, in 1941, the time dilatation effect was observed by Rossi and Hall with their famous experiment involving extremely high-speed particles (muons) coming from the cosmos. This experiment is presented hereafter.

Since then, many experiments involving extremely high speeds were performed, and all confirmed the accuracy of the relativistic laws. In particular, electrons could reach the speed of 0.999 999 999 983 c at the CERN particle accelerator.

2.3.1.1 *The Rossi & Hall Experiment at Mount Washington*

Muons (mu-mesons) are very light particles (207 times the electron mass), which are the result of reactions between particles of the cosmic rays and the high atmosphere. Muons approach the Earth surface at the extremely high speed of 0.992 c, but as they are unstable, they disintegrate with the half-life of 1.53 μs, meaning that a population of muons is divided by 2 every 1.53 μs.

A muon detector was placed at the Mount Washington with the altitude of 1910 m. This detector counted 563 ± 10 muons per hour. A second detector was put at a much lower point of 3 m of altitude, and it counted 408 ± 9 muons per hour.

We can calculate the time taken by the muons to go from the high detector to the low one, and this will enable us to calculate the number of muons that disintegrate during this trip. We will then deduce the theoretical number of muons that reach the low altitude of 3 m.

Using classical physics, the calculation gives much fewer muons than what was observed: indeed, the travel duration of the muon is: (1910–3) m/0.992.c = 6.408 μs, which is slightly more than four times their disintegration period. We then expect to see less than $563/(2^4) = 35.18$ particles per hour versus 408 observed.

When applying the laws of Relativity, this problem was solved: the disintegration period of the muon is indeed to be considered in the frame of the muon, meaning with the proper time of the muon. Consequently, the duration of 6.408 μs which is measured on the Earth must be converted into the duration observed by the muon and measured with its proper time. The time dilatation law says that this travel time considered by the muon is

seen dilated on Earth by the γ factor, which is in this case: $\sqrt[2]{\dfrac{1}{1-0.992^2}} = 7.92$.

Consequently, the travel duration seen by the muon with its proper time is almost eight times less: $6.408/7.92 = 0.809$ μs.

This travel duration corresponds to $0.809/1.53 = 0.529$ times the disintegration period, which leads to the following number of muons at the low altitude of 3 m: $563.(1/2)^{0.563} = 390$ muons, which is close to what was observed (~5% difference) ∎

2.3.1.2 Theoretical Aspects

The new theory solved the problems raised by Maxwell's equations, which was the initial objective pursued by A. Einstein, H. Poincaré and A. Lorentz. With the Lorentz transformation, Maxwell's equations have the same form in all inertial frames. Moreover, Relativity showed that the electric and magnetic fields actually are a single entity, as we will see in Volume II Section 5.1.

Furthermore, Einstein showed that Relativity concerns the whole of physics since it broke the Newtonian paradigm of an absolute frame. We will see that Relativity subsequently revealed new laws especially regarding the energy, represented by the famous one, $E = mc^2$. We will also see that Special Relativity has elegantly unified several different laws, most significantly the fundamental conservation laws regarding the momentum and the energy of closed systems.

However, one domain resisted the new paradigm: gravitation. This phenomenon was thought, according to Newton, to exert instantaneous distant interactions, contradicting an important finding of Special Relativity that nothing can surpass light speed. However, after 10 years of intense and complex work, Einstein solved this problem with his theory of General Relativity.

2.3.2 An Important Application: The Doppler Effect

The Doppler effect is important due to its numerous applications in various domains, in particular cosmology, car speed measurements, medicine, etc. The light speed independence of its source (the "affront to the common sense") raised a serious problem to the Doppler effect with respect to the principle of Relativity, in a similar fashion to Maxwell's equations. Hence, we will first calculate the Doppler effect with the classical kinematics law and see this problem. We will then show the way Relativity solved it.

The Doppler effect consists of the following:
A source S periodically emits light pulses; the time duration between two pulse emissions is called the emitter period, denoted by ΔT_E. This source S is moving away from an observer located at the point O, with a constant speed v relative to O and along the straight line OS. The observer receives these pulses, and the time duration between two pulse receptions is called the receiver period, denoted by ΔT_0 (for the observer) (Figure 2.9).

We want to calculate ΔT_0 knowing ΔT_E. We will first use the classical laws, and we

O•------------------------- ((((((•S→•S'

FIGURE 2.9
A moving source S periodically emits signals toward an observer O.

will successively make the calculations from the observer's frame, and then from the source's frame. We will see that these results are different, which is not acceptable as it contradicts the principle of Relativity.

We will then consider the same cases, but applying the Relativity principles, and the problem will be solved.

2.3.2.1 Applying Classical Laws Contradicts the Principle of Relativity

2.3.2.1.1 Case Where the Source Is Moving, Whereas the Observer Is Fixed

Let's place ourselves in the shoes of the observer O; we reason as follows:
Let's call T_1 the time at which the 1st pulse is emitted, and T_2 the time at which the 2nd pulse is emitted. The emission period is: $\Delta T_E = T_2 - T_1$.

The point O receives the first pulse at the time T_1^o and the second one at T_2^o. The period received in O is then: $\Delta T_0 = T_2^o - T_1^o$.

Each pulse is going at the speed c, independently of the speed of its source (as a consequence of Maxwell equations in the context of no ether). Hence, the time taken by the 1st pulse to go from S to O is: SO/c. Consequently:

$$T_1^o = T_1 + SO/c.$$

The second pulse is emitted when the source is at the point S', thus:

$$SS' = v.(T_2 - T_1) = v. \Delta T_E.$$

The time taken by the 2nd pulse to go from S' to O is: S'O/c; hence: $T_2^o = T_2 + S'O/c$. Then:

$$\Delta T_0 = T_2^o - T_1^o = (T_2 + S'O/c) - (T_1 + SO/c) = (T_2 - T_1) + (S'O - SO)/c = \Delta T_E + SS'/c.$$

Having seen that SS' = v.ΔT_E, we finally obtain:

$$\Delta T_0 = \Delta T_E (1 + v/c) = \Delta T_E (1 + \beta). \tag{2.12}$$

We have replaced v/c with β, which is usually done in Relativity.

2.3.2.1.2 Case Where the Observer Is Moving, Whereas the Source Is Fixed

According to the inertial frames equivalence postulate, we can consider the same scenario, but with the source fixed, and the point O moving at the speed $-v$ in the opposite direction of the first case. Let's then calculate in this context the relation between the emitter period and the receiver one (Figure 2.10):

The 1st pulse emitted by S at the time T_1 is received by O at the time T_1^o;

O'←O•----------------------- ((((((•S

FIGURE 2.10
A "fixed" source S periodically emits signals toward a moving observer.

thus: $T_1^o = T_1 + SO/c$.

The observer receives the 2nd pulse at the time T_2^o. At that time, he has covered the distance OO' which is: OO' = v ($T_2^o - T_1^o$) = v ΔT_0. This 2nd pulse was emitted by the source S at the time $T_2 = T_2^o - SO'/c$. We then have:

$$\Delta T_E = T_2 - T_1 = (T_2^o - SO'/c) - (T_1^o - SO/c).$$

Then: $\Delta T_E = \Delta T_0 - (SO' - SO)/c = \Delta T_0 - OO'/c = \Delta T_0 (1 - v/c)$.

So finally:

$$\Delta T_0 = \frac{\Delta T_E}{1 - v/c} = \frac{\Delta T_E}{1 - \beta}. \tag{2.13}$$

Conclusion: The equations 2.12 and 2.13 are different, giving always different results, which contradicts the inertial reference frames equivalence postulate. The differences are very small as long as the speed v is small relative to c, but they are huge when v is close to c. This situation is unacceptable because, for instance, the observer O can have an electronic device which displays the frequency received and it will display only one number which can be seen by anyone in any frame.

Note that this contradiction appears because we took the light speed invariance postulate. Otherwise, there was no contradiction in classical physics.

**

2.3.2.2 The Doppler Effect Revisited in Light of Relativity

We notice that there are two different inertial frames: the frame K, where the observer O is fixed, and the frame K' of the emitting source S. Relativity showed that times and distances are relative to a frame. This means that the times mentioned in relations in equations 2.12 and 2.13 are not directly comparable. To compare them, we will identify the intrinsic elements since these are the same in all frames, such as the proper times and the proper distances. We will then take advantage of the relativistic rules: time dilatation and distance contraction.

Let's then identify in both relations which are the proper times:

In the relation 2.12, ΔT_0 is a proper time because the pulses are received at the same fixed point O of K. Hence, ΔT_0 will be denoted by $\Delta \tau_0$ to mark that it is a proper time duration.

The relation 2.12 thus becomes:

$$\Delta \tau_0 = \Delta T_E \left(1 + \beta\right). \tag{2.14}$$

In the relation 2.13, ΔT_E is a proper time because the pulses are emitted in the same fixed point S of K'. Hence, ΔT_E will be denoted by $\Delta \tau_e$ to mark that it is a proper time. Likewise, we mark $\Delta T'_O$ the period received in O since this period is considered in the frame K'. The relation 2.13 then becomes:

$$\Delta T'_O = \frac{\Delta \tau_e}{1 - \beta}. \tag{2.15}$$

Let's now apply the relativistic time dilatation law:

In relation 2.14, we have: $\Delta T_E = \gamma \Delta \tau_e$, and in relation 2.15: $\Delta T'_O = \gamma \Delta \tau_0$. Hence, equation 2.14 becomes:

$$\Delta \tau_0 = \gamma \Delta \tau_e \left(1 + \beta\right). \tag{2.16}$$

Similarly, equation 2.15 becomes:

$$\gamma \Delta \tau_0 = \frac{\Delta \tau_e}{1 - \beta}. \tag{2.17}$$

We will show that there is no discrepancy between relations 2.16 and 2.17: for this, let's multiply the numerator and the denominator of equation 2.17 by $(1+\beta)$; we obtain: $\gamma \Delta\tau_0 = \dfrac{\Delta\tau_e (1+\beta)}{1-\beta^2}$; then replacing $\dfrac{1}{1-\beta^2}$ with γ^2, we obtain:

$\gamma \Delta\tau_0 = \gamma^2 \Delta\tau_e (1+\beta)$, which finally gives:

$\Delta\tau_0 = \gamma \Delta\tau_e (1+\beta)$ ∎ This relation is identical to 2.16.

Conclusion: In Relativity, the Doppler effect complies with the inertial frames equivalence postulate. What is more, we can see that the relativistic period received by the observer in O is equal to the one calculated classically in the receiver's frame multiplied by γ.

Note that in the relation 2.16, if we replace γ with $\dfrac{1}{\sqrt{1+\beta}\sqrt{1-\beta}}$, we find:

$$\Delta\tau_0 = \Delta\tau_e \frac{\sqrt{1+\beta}}{\sqrt{1-\beta}} \qquad \blacksquare (2.18)$$

2.3.2.3 Complements and Remarks on the Doppler Effect

The relation between the proper frequency of an emitter, F_e, and the observer's proper frequency received, F_o, can be deduced from relation 2.18 since: $F_e = 1/\Delta\tau_0$ and $F_o = 1/\Delta\tau_0$. Then:

$$\mathbf{F_o = F_e} \frac{\sqrt{1-\beta}}{\sqrt{1+\beta}}.$$

The frequency you receive from an emitter that is moving away from you is lower than the proper frequency of that emitter, and the relativistic effect amplifies the classical Doppler effect by the γ factor.

Likewise, the relation between the wavelengths is obtained with the general relation between a wavelength λ and a period or a frequency: $\lambda = c\,\Delta\tau = c/f$.

Remark 1: The Doppler effect applies not only in the air, but also in various fluids, hence its numerous applications in various domains, including acoustics (e.g., the noise of a moving car).

Remark 2: In the general case where the emitter velocity is not parallel to the line observer-emitter, but makes an angle θ relative to this line, then we should take into account the projection of the velocity to this line and the Doppler formula becomes:

$$\Delta\tau_0 = \gamma \Delta\tau_e (1 + \beta\cos\theta).$$

Remark 3: In astronomy and cosmology, the Doppler effect plays a key role: E. Hubble noticed in 1925 that the further a galaxy, the more its light spectrum was shifted toward the red (hence the name "redshift"). This meant that the further the galaxies, the faster they move away from us. Looking backward in time, this suggested that the universe was very concentrated, which supported the Big Bang theory[4]. Relativity shows that the Doppler effect is significantly amplified concerning the most distant galaxies.

Remark 4: Another geometrical demonstration of the relativistic Doppler effect is given with Problem 2.12 using the Minkowski diagram method.

Remark 5: Historically, the German scientist W. Voigt, who did not believe in the ether, found in 1887 a mathematical solution to the Doppler problem with respect to the principle of Relativity. He thus found a solution which was very close to the Lorentz transformation, but its physical meaning was not clear.

2.4 Minkowski Diagrams

Minkowski diagrams are very useful to represent various scenarios in a simple geometrical manner, and to solve many practical problems. We will successively see Minkowski diagrams with one frame, and then two frames.

2.4.1 Minkowski Diagrams with One Frame

The starting point is to represent events with only their coordinates along the OX and OT axes in an inertial frame K, taking advantage of the fact that there is no Relativity along the transverse directions (OY and OZ).

The second idea is to express the time in a time unit (e.g., the second) multiplied by c, meaning $t \rightarrow ct$. This enables us to visualize very fast projectiles, even light pulses. There is another theoretical reason for choosing c as we will further see with the time-distance equivalence (Section 3.4).

Consider the following scenario: in the frame K, at the time $t = 0$, the point O emits a light pulse which propagates along the OX axis. On this axis, there is a fixed point A in K such that OA = 1 meter. We denote by A* the event "the pulse is in front of the point A", and we want to represent this event in a Minkowski diagram:

This event is represented in the Minkowski diagram below by the point A* whose abscissa is 1. Regarding its ordinate, let's denote t_a the time in K at which the event A* takes place. The ordinate of A* in the Minkowski diagram is: $c.t_a$. It remains to calculate t_a: since the pulse goes at the velocity c, we have: $d = c.t$, meaning for the event A* that: $1 = c.t_a$. Hence, the ordinate of A* in the Minkowski diagram is 1, as represented in Figure 2.11:

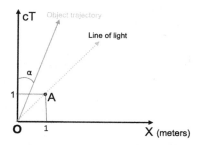

FIGURE 2.11
Minkowski diagram with one frame.

More generally, the trajectory of this light pulse emitted from O is the diagonal line, called the "line of light", because all its events satisfy the relation: ct = x.

A real projectile leaving the point O at the time t = 0 and moving at a constant speed v along the OX axis has a trajectory which is a straight line in the Minkowski diagram, because all events of this trajectory satisfy the relation: $t = x/v$; then: $ct = x.c/v = x. \dfrac{1}{\beta}$.

The slope of this object trajectory is thus: $1/\beta$. It is always greater than 1 because the speed v must be below c, as we will further see. The object trajectory is thus a line which is always above the line of light; moreover, the faster the projectile, the greater the angle α of its trajectory with the OcT axis, since: **tg(α) = v/c = β**. Hence, when the object speed tends to c, its trajectory tends to be the line of light.

2.4.1.1 Example of Minkowski Diagram with One Frame

This example shows the difference between the current position of a rocket or a star and its observed position.

A photon emitted by a star or a very distant and fast rocket can take several years to reach us on Earth. Hence, during the travel duration of this photon, the rocket (or the star) may have moved very significantly. Let's assume that the rocket moves away from us at the constant speed v along a straight line passing by the Earth.

The frame K is the Earth frame. An observer stays at the point O, and observes a rocket going further and further from the Earth at the constant speed v. For the sake of simplicity, we set the origin-event of K to be when the rocket left the Earth. This observer receives a photon emitted by the rocket at the time To of K. The event corresponding to the reception of the photon thus has the coordinates in K: (0, To). However, this photon was emitted long before, at the time denoted by Te in K. We want to determine Te and the distance rocket–the Earth at the time Te (Figure 2.12):

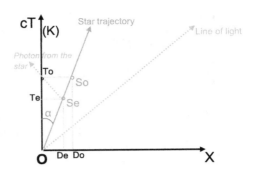

FIGURE 2.12
Photons received from a moving star °

In the above Minkowski diagram, the rocket (or star) trajectory is along the straight line passing by O and having the angle α such that: $tg(\alpha) = v/c = \beta$.

The trajectory followed by the photon is perpendicular to the line of light, its equation being: $ct = -x + cT_o$. Hence, this photon was emitted at the point Se which is at the intersection of the rocket trajectory and the photon trajectory. This point Se represents the event "the rocket emits the photon which will be received by the observer at To".

The ordinate of Se is the time in K, denoted by Te, at which the photon was emitted, multiplied by c. The abscissa of Se gives the distance rocket-Earth denoted by De at the time Te (in K). Thus, this Minkowski diagram answered our questions.

Additionally in this case, the above Minkowski diagram enables us to find geometrically the relation between To and Te as we will see:

We notice that the length of the segment $cT_o \_ cT_e$ is equal to the one of the segment $cT_e \_ S_e$, because the angle $(cT_e\ S_e\ cT_o)$ is 45°. We also have:

$cT_e \_ S_e = T_e\ tg\alpha = cT_e\ \beta$. Hence: $T_o - T_e = T_e\ \beta$, so finally: $T_e = T_o/(1+\beta)$ ∎

The relation between the distances of the rocket to the Earth is obtained knowing that the rocket velocity is v, and at $t = 0$, the distance was null. Hence by multiplying both terms by v, we obtain: $D_e = D_o/(1+\beta)$.

Note that in the above diagram, it would be convenient to choose the light-year as distance unit in the OX axis, and the year in the OcT axis.

Remark: This scenario can be used to calculate the Doppler formula, as shown in Problem 2.12.

2.4.2 Minkowski Diagrams with Two Combined Frames

Two inertial frames K and K′ share a common origin-event O*, with K′ moving at the velocity v relative to K. The Minkowski diagram (Figure 2.13) with two frames functions as follows:

The frame K is represented by an orthonormal frame of Minkowski, as previously, but not the frame K′; however, both frames share their common

bisector between (OX, OcT) and (O'X', O'cT'), which is also their common line
of light: indeed, a photon leaving O and O' at the time t = t' = 0, and then moving along the OX and O'X' axes, always satisfies: x = ct in K, and x' = ct' in K'.

The point O' of K' moves at the speed v relative to K along the OX axis of K;
hence, its trajectory is a straight line as seen previously, and its slope is such
that tg(α) = v/c = β. Besides in K', the trajectory of the point O' is the OcT' axis
since its abscissa is always null.

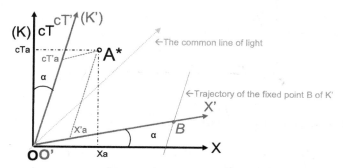

FIGURE 2.13
Minkowski diagram combining two frames.

Consider a random event A* whose coordinates, respectively, are (Xa, cTa) in
K and (X'a, T'a) in K'. The beauty of the Minkowski diagram with two frames
is that the event A* is represented by the same point A* in both Minkowski
frames: its coordinates, respectively, are (Xa, cTa) in the Minkowski frame K,
and (X'a, cT'a) in the Minkowski frame K'. This important property is demonstrated in Volume II Section 2.3.1.

A scenario is described by a set of events. In such Minkowski diagrams,
the same sets of points represent the same scenario in K and K'. In order to
better visualize scenarios in both frames, it is recommended that fixed points
in one frame are represented with their trajectories (worldlines) since they
are seen moving in the other frame. This is done with the point B of K' in
Figure 2.13.

2.4.2.1 The Time Dilatation and the Different Lengths of the Unit Frame Vectors in K and K'

Consider an event A* which is: "the point O' of K' emits a flash at the time
ct' = 1 in K'". The point A* thus has the ordinate 1 along the cT', as shown in
the diagram below. According to the diagram rule, this event has the ordinate
cTa in K, which is the intersection of the OcT axis and the parallel to OX passing by A*. We can see that the length of the segment O-cTa is smaller than the
length O-A*, which seems to contradict the time dilatation law. Indeed, the
length of O-cTa should be OA*.γ, whereas it is OA*.cosα. Actually, this is not
a contradiction since it has an explanation: the unit frame vectors of K don't

have the same length as those of K'. Those of K are indeed smaller than those of K' by the ratio $\gamma/\cos\alpha$ (Figure 2.14).

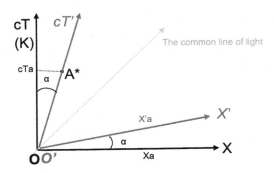

FIGURE 2.14
Time dilatation seen with the Minkowski diagram.

For instance, if $v = 0.4$ c, then: $\alpha = \text{atan}(0.4) = 21.8°$, $\gamma = 1.0911$, $\cos\alpha = 0.928$. As a result, the frame unit vectors of K are shorter than those of K' by the ratio: $1.0911/0.928 = 1.175$.

This drawback does not exist in another type of diagram, the Minkowski-Loedel diagram, which is presented in Volume II Section 2.3.2. However, Minkowski diagrams are still very commonly used and relatively easy to manipulate. Hence, these will be used in Volume I.

2.5 Mathematical Supplements

2.5.1 Demonstration: The Invariance Under Translation Implies the Linearity of Φ

Having shown in Section 2.3.1 that the image by Φ of any space-time separation Four-Vector is invariant under translation of this separation vector, we will demonstrate the linearity of the transformation function Φ. For this, we will proceed in three steps:

Step 1: We will show that for any space-time separation vector seen in K by the vector $\vec{u}$, and for any number k, we have: $\Phi(k.\,\vec{u}) = k.\Phi(\vec{u})$.

Step 2: We will show that for any couple of space-time separation vectors seen in K, $\vec{u}$ and $\vec{v}$, we have: $\Phi(\vec{u} + \vec{v}) = \Phi(\vec{u}) + \Phi(\vec{v})$.

Step 3: We will demonstrate the linearity of Φ: $\Phi(x.\vec{u} + y.\vec{v}) = x.\Phi(\vec{u}) + y.\Phi(\vec{v})$.

*

Step 1: We will first show this result with an example: $k = 3$. Then, we will generalize to any value of k.

In the frame K, the events A* and B* are seen at the points A and B, defining the vector $\vec{u} = \overrightarrow{AB}$.

There is a point M in K such that: $\overrightarrow{OM} = \vec{u}$. We can then define the points N and P such that: $\overrightarrow{MN} = \vec{u}$ and $\overrightarrow{NP} = \vec{u}$. We then have in K: $\overrightarrow{OP} = 3.\vec{u}$. These three points M, N and P represent the events M*, N* and P* seen in K (Figure 2.15).

In K', the same events A* and B* are seen at the points A' and B', defining the vector $\vec{u'} = \overrightarrow{A'B'}$.

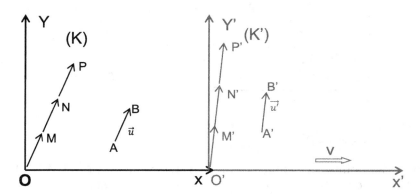

FIGURE 2.15
The vector 3 $\vec{u}$ is equal to $\overrightarrow{OP}$.

We have: $\vec{u'} = \Phi(\vec{u})$ by definition of Φ. Let's denote by M', N' and P' the images in K' of the events M*, N* and P*. As $\overrightarrow{OM} = \vec{u}$, the invariance under translation implies: $\overrightarrow{O'M'} = \vec{u'}$. The same applies to $\overrightarrow{MN}$ and $\overrightarrow{NP}$: since both are equal to $\vec{u}$, their images are equal to $\vec{u'}$; consequently: $\overrightarrow{M'N'} = \overrightarrow{N'P'} = \vec{u'}$.

Then as P' is the image of P, we have: $\overrightarrow{O'P'} = \Phi(\overrightarrow{OP})$, so finally: $\Phi(3. \vec{u}) = 3.\vec{u'}$.

Generalization: This reasoning obviously applies to any value of k, provided that it is any integer. Let's then address the case where k is not an integer, for instance, k = 3.45. We will show that: $\Phi(3.45. \vec{u}) = 3.45 \Phi(\vec{u})$. For this, we will first show that:

$100 \times \Phi(3.45. \vec{u}) = 100 \times 3.45 \Phi(\vec{u})$, and then by dividing all by 100, we will obtain the result.

According to Step 1, the left part, $100 \times \Phi(3.45. \vec{u})$, is equal to:
$\Phi(100 \times 3.45. \vec{u}) = \Phi(345. \vec{u})$.

The right part, $100 \times 3.45 \Phi(\vec{u})$, is equal to: $345 \Phi(\vec{u}) = \Phi(345. \vec{u})$.

This reasoning is valid for any value of k provided that it is a rational number (k = p/q with p and q being integers; then instead of multiplying by 100, we multiply by q). We will assume that the function Φ is continuous; then k can be a real number because there always exists a rational number which is as close as we want to any real number; subsequently, the image of this rational number will be as close as we want of the image of the real number.

Step 2: Consider two space-time separation vectors which are, respectively, seen in K by the vectors $\vec{u}$ and $\vec{v}$, and in K′ by the vectors $\vec{u'}$ and $\vec{v'}$. In K, there exist two points M and N such that $\overrightarrow{OM} = \vec{u}$ and $\overrightarrow{ON} = \vec{v}$. We can then define in K the point P such that $\overrightarrow{MP} = \vec{v}$, and we thus have: $\overrightarrow{OP} = \vec{u} + \vec{v}$ (Figure 2.16).

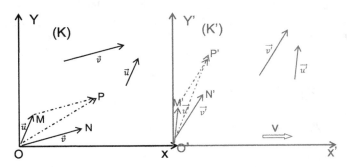

FIGURE 2.16
The sum $\vec{u} + \vec{v}$ is equal to $\overrightarrow{OP}$.

In K, the points M, N and P represent the events M*, N* and P*; these events are seen in K′ at the points M′, N′ and P′. In K′, we have: $\Phi(\vec{u}) = \vec{u'}$ and $\Phi(\vec{v}) = \vec{v'}$. Then, thanks to the invariance under translation, we have:

$\vec{u'} = \overrightarrow{O'M'} = \Phi(\overrightarrow{OM}); \vec{v'} = \overrightarrow{O'N'} = \Phi(\overrightarrow{ON})$.

Also: $\overrightarrow{M'P'} = \Phi(\overrightarrow{MP}) = \Phi(\vec{v}) = \vec{v'}$. Then:

$\overrightarrow{O'P'} = \overrightarrow{O'M'} + \overrightarrow{M'P'} = \vec{u'} + \vec{v'} = \Phi(\vec{u}) + \Phi(\vec{v})$.

P′ being the image of P: $\overrightarrow{O'P'} = \Phi(\overrightarrow{OP}) = \Phi(\vec{u} + \vec{v})$. Hence finally:

$\Phi(\vec{u} + \vec{v}) = \Phi(\vec{u}) + \Phi(\vec{v})$.

Step 3: Consider the vectors: $\vec{U} = x.\vec{u}$ and $\vec{V} = y.\vec{v}$. From what precedes: $\Phi(\vec{U} + \vec{V}) = \Phi(\vec{U}) + \Phi(\vec{V})$. We then have:

$\Phi(x.\vec{u} + y.\vec{v}) = \Phi(\vec{U}) + \Phi(\vec{V}) = \Phi(x.\vec{u}) + \Phi(y.\vec{v}) = x.\Phi(\vec{u}) + y.\Phi(\vec{v})$ ∎

2.6 Questions and Problems

Problem 2.1: Appearance of a Cube seen from a Fast Space Ship

Imagine that you are in an extremely fast space ship, represented by the frame K′. You pass near a space station that has a cubic shape in its rest frame; your velocity, v, relative to this cubic station is very high, perhaps 0.9 c (Figure 2.17).

At the time t'_0 of K′, you are in front of the middle of the station. What shape does the station that you see have: Is it: a), b), c) or d)? We consider for the sake of simplicity that the photons received at your eye at t'_0 have traveled the same distance whatever the point of the station from which they were emitted.

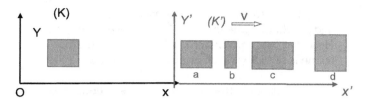

FIGURE 2.17
How is a cube in K seen in K'?

Remark: The same case is treated in Volume II, with Problem 2.13 where this simplifying assumption is not made.

Problem 2.2: Some useful relations

Show that: $\gamma^2(1-\beta^2) = 1$; then: $\beta^2 = 1 - \dfrac{1}{\gamma^2}$; then: $\beta.\gamma = \sqrt{\gamma^2 - 1}$.

Show that the determinant of the Lorentz matrix equals to 1.

Problem 2.3: Surprising Effects when Traveling in the Cosmos – Part 1

Bob is undertaking a journey into space, whereas Alice stays on Earth. Bob and Alice set their clocks at zero when Bob leaves Alice. Bob's rocket goes at the very fast speed of 0.943 c along a straight line; this speed corresponds to the gamma value of 3.00. Bob's clock shows 10 years when he arrives at the distance of the first planet, named Proxima. We will consider that both Alice's and Bob's frames are inertial, despite the circular movements of the Earth and the acceleration of Bob's rocket.

Q1: For Alice, how long did Bob's journey take to reach Proxima?

Q2: How can we explain the dissymmetry between these travel durations (10 versus 30 years)?

Q3: At the year 30 of Alice's clock, she initiates a phone call in order to congratulate Bob for arriving at the level of Proxima. Let's now place ourselves in Bob's rocket frame. What time is it for Bob when Alice initiates her call? Please comment.

Q4: How can we explain the discrepancy between these results: Bob spent 10 years reaching Proxima, whereas Alice's call occurs 90 years after his departure?

Q5: We assume Proxima is fixed relative to the Earth. What is the distance Earth–Proxima seen by Bob? Then, seen from Alice? Which one is the proper length Earth–Proxima?

   – Further comments: If Bob's rocket accelerates and reaches, for instance, the speed 0.9999999 c, then the universe will be seen much smaller by Bob. Hence with such extreme speeds, it can be possible to travel from one end to the other of the universe (but provided that the universe does not expand rapidly). One may be induced to apply these considerations to the photon, and deduce that for the photon, the distance of the universe is null. However, we cannot make this deduction since the photon's frame is not inertial (cf. more in Section 5.4).

More questions/exercises on this Bob and Alice scenario are in Volume II Problem 2.3.

Problem 2.4: Scenario showing the Absurdity of Writing $t' = \gamma_v.t$ in the General case

Consider two inertial frames K and K'. We will show that writing $t = \gamma_v.t'$ leads to an absurdity. For this, let's have two events E* and F* which are simultaneous in K': $t'_E = t'_F$. Show that applying the relation $t = \gamma_v.t'$ to these events leads to an absurdity.

Problem 2.5: Show that the Transformation of a Straight Line in K is a Straight Line in K'. Use only the Linearity of the transformation Φ.

Consider a set of events which are seen along a straight line in an inertial frame K. Show that this set of events is also seen along a straight line in any inertial frames, by using the linearity of the transformation function Φ. Hint: choose one point of the straight line of K as the common origin-event of K and K'.

Problem 2.6: Complement on the Rossi & Hall Experiment at Mount Washington

We take the same scenario statement as in Section 2.3.1.1, and we want to calculate the number of muons that arrive at the sea level by another argument: imagine that you are fixed with a muon; you see the Earth's surface arriving toward you very fast:

Q1: At which speed is the Earth's surface coming toward you?

Q2: When you are at the level of Mount Washington, what is the distance between you and the Earth's surface, as measured by you in the same frame as the muon?

Q3: How long do you have before the crash? Compare this time with the muon disintegration period and conclude about the number of muons arriving at the surface level.

Problem 2.7: Check that the Lorentz Transformation Implements the Time Dilatation Law

In an inertial frame K, let's have a fixed clock at the point A of the OX axis, its abscissas being a. Consider the two following events E* and F*: E* is "the clock displays the time t1"; and F* is: "the clock displays the time t2". In another frame K' moving at the speed v relative to K along the OX axis, these same events E* and F* have the following coordinates: E' (m', t'1) and F' (n', t'2).

Q1: Which is a proper time: $(t2 - t1)$ or $(t'2 - t'1)$?

Q2: These two events define the vector E^*F^*, which is denoted by EF in K and E'F' in K'. Express the coordinates of EF in K, and those of E'F' in K'.

Q3: Express the relation between EF and E'F' using the matrix format.

Q4: Expand the line of this matrix relation which concerns the time, and then conclude.

Problem 2.8: Check that the Chronological Order between two Co-Located Events is Invariant

This is a supplement to the previous exercise: consider the case where the event E* is seen occurring before the event F* in the frame K where they are seen co-located. This means that the event E* can be the cause of the event F*. In another frame K', these events are not co-located, but it would be absurd if in K', the event F* was seen prior to the event E*. Show that such absurdity cannot occur, in other words that the chronological order between two co-located events is unchanged in other frames.

Problem 2.9: Check that the Lorentz Transformation Implements the Distance Contraction Law

Consider a ruler of length L in the inertial frame K; this ruler is fixed and parallel to the X axis, with its two ends being the points A and B. The abscissas of A and B are, respectively, a and b in K.

This ruler is seen in another inertial frame K', moving at the speed v relative to K along the OX and O'X' axes. The extremities of this ruler are seen in the moving points A' and B' of K', and the length of the ruler in K' is denoted by L'.

Q1: Which is the proper length of the ruler: L or L'?

Q2: An observer in K' wants to assess the length A'B': Can he assess $L' = b'-a'$ by measuring the abscissa a' of A', at a different time in K' than the abscissa b' of B'?

Q3: Let's call t' the time of K' at which the observer in K' measures a' and b'. This defines two events: event A*: "in K', the point A' is at the abscissa a' at the time t'''; and event B*: "in K', the point B' is at the abscissa b' at the time t'''. Express the coordinates in K' of the vector A*B*, denoted by A'B'.

Q4: Are the events A* and B* simultaneous in K?

Q5: We, respectively, denote by t1 and t2 the times in K at which the events A* and B* occur. Express the coordinates in K of the space-time vector A*B*, denoted by AB.

Q6: Express the relation giving (A'B') as a function of (AB) using the matrix format.

Q7: Expand the line corresponding to the space part, and conclude: Is the length contraction law respected?

Problem 2.10: Axiomatic Questions on Our demonstration of the Lorentz Transformation

Our demonstration of the Lorentz transformation used two scenarios, but neither involved the length contraction law, nor even the light speed invariance. At first glance, this may seem abnormal: Does it mean that these laws (or postulates) are unnecessary to demonstrate the Lorentz transformation?

Problem 2.11: Complement to Einstein's Train Scenario: Typical Time Difference between the two Pulses seen from the Train

This problem refers to Einstein's train scenario described in Section 1.1.2.1. The train has a proper length of 200 m and goes at the high speed of 300 km/h. On the platform, the points A and B simultaneously emit light pulses.

Question: Seen from the train, what is the time difference between the two pulse emissions?

The following notation will be used: we denote by K the frame of the platform, and K' the one of the train. The two light pulses simultaneously emitted from the points A and B of the platform define two events E* and F*. For the sake of simplicity, K and K' have a common origin-event which is E* (light pulse emission from A).

Problem 2.12: Minkowski Diagram with one Frame: The Relativistic Doppler Effect

We will reuse the scenario of Section 2.4.1.1, and we replace the star with an object leaving the Earth at the speed v. This object has an emitter which periodically emits light pulses toward the Earth, the first one occurring when the object leaves Earth, which is at $T = 0$ (Earth time); the second one is received on Earth at To. The observed period on Earth is thus To.

This second pulse has been emitted by the object at the time Te of K (Earth frame), which thus corresponds to the emitter period.

Q1: Is Te the proper period of the emitter?

Q2: What is the relation between Te and To? Use a Minkowski diagram to determine it.

Q3: If Te is not a proper period, we will note τ_e the proper period of the emitter. What is the relation between Te and τ_e? Then compare with the Doppler formula.

Problem 2.13: Minkowski Diagram with 2 frames: Einstein's Train Scenario

Let's take again the Einstein's famous train scenario: the train frame is K', and the platform frame is K. We represent this scenario with a Minkowski diagram with two frames, and both frames share the common origin-event: "the back end of the train is in front of O and O'".

We mark a point B on the platform, such that the front end of the train is in front of B at the time t = 0 of K.

Q1: Does OB represent the proper length of the train?

Q 2A: We have placed two lasers on the platform, one at O and one at B. These lasers simultaneously emit a light pulse at the time t = 0 of the ground. The laser of O emits its pulse in the same direction as the train velocity, whereas the laser in B emits its pulse in the opposite direction. Represent the events O* and B* corresponding to the flash emissions in O and B at t = 0.

Q2B: Represent the trajectories of these two flashes.

Q3A: We call C* the event: "both flashes coincide". Represent this event, and show its abscissa Xc in K and X'c in K'.

Q3B: Express Xc as a function of Xb.

Q4: We want to know if the event C* is seen in the middle of the train. For this, we call F the point tracking the front end of the train. Represent the trajectory (worldline) of the point F.

Q5A: We mark F* the point corresponding to the location of F at the time t' = 0. What does the distance O'F* represent and why?

Q5B: The abscissa of F* in K is Xf. Does Xf represent the train length seen from K?

Q5C: If not, what is the length of the train seen from the platform?

Q6: Does the event C* (flash coincidence) occur in the middle of the train? If not, does it occur closer to the front or to the back of the train? Please comment.

Q7: At the time t' = 0, the front end of the train emits a flash. Show in the diagram the time at which this flash occurs in K.

Q8: In the train, does the event B* occur before or after the flash emission in O?

Problem 2.14: Time Dilatation and De-synchronization Effect

There is a clock at every window of a fast train, and all are synchronized with the train frame K'. The proper distance between each window of the train is 3 m. The first window is denoted by A, the next one by B, and so forth. An observer is staying on the ground at a point O, and sees the train passing in front of him. The Earth frame, denoted by K, is assumed to be inertial. We set the common origin-event O* of K and K' to be when the clock A of the train is in front of the observer. Thus, this clock displays the time zero at the event O*.

Q1A: Express the time of the clock B seen from the ground at t = 0, and denoted by Tb. The train speed is denoted by v, and the proper distance between the clocks A and B is denoted by AB.

Q1B: At the time 0 on the ground, does the observer see the clock B delaying or advancing compared to the clock in A?

Q1C: The train is moving at 300 km/h. What is the difference between the times of the clocks A and B seen from the ground at $t = 0$?

Q2: The next window after B has a clock C. (B is in the middle of A and C.) Compare the time difference seen from the ground between the clock C and the clock B, and the one between the clock B and the clock A.

Q3: Make a Minkowski diagram of the above scenario, and answer Question 2.

Q4: What would be the train speed if we want the time difference between two consecutive windows to be 1 second as seen from the ground? (You may make first-order approximations.)

Problem 2.15: The Doppler Effect

Some light has been received from a celestial body that is going away from the Earth at the speed of 0.85 c. One line of its spectrum corresponds to the wavelength of $3{,}400 \cdot 10^{-10}$ m. What was the wavelength of this line at its emission? What was its frequency emitted; and what is its frequency received on Earth?

Problem 2.16: Length Contraction Demonstration Presented as an Exercise

A train is moving at the speed v and passes in front of a station. We place an observer at the point A on the platform with a clock: he marks the time t1 of his clock when the front end of the train arrives in front of him, and then he marks the time t2 when the back end of the train is in front of him (Figure 2.18).

FIGURE 2.18
Length of a moving train.

Q1: What is the train length, denoted by L, seen from the platform?

Q2: Let's now imagine that we are inside the train, which is an inertial frame noted K': we place one observer at the front, and another one at the back. They both see the platform and the point A moving backward at the speed –v. The observer at the front end marks the time t'1 (of K') when he is in front of the point A; then the one at the back marks the time t'2 (of K') when he is in front of the point A. What is the train length seen from the train? We will denote by L' this length.

Q3: Which is the proper length of the train? This proper length will be denoted by L_p.

We will now establish the relationship between L and L_p, and to this purpose, we will seek the relation between (t'2 – t'1) and (t2 – t1). We will call E the event: "the front end of the train is in front of the point A", and F the event "the back end of the train is in front of the point A".

Q4: Is (t'2 – t'1) or (t2 – t1) a proper time duration?

Q5: What is the relation between (t'2 – t'1) and (t2 – t1)?

Q6: What is the relation between L and L_p? What do you conclude?

Problem 2.17: Simultaneous Light Pulses

This problem is a continuation of Problem 1.2. You are in a train, your friend is on the platform at a point marked O, and there is a tunnel 3,000 m ahead. You both have a laser, and when you are exactly in front of your friend, both of you simultaneously emit a light pulse toward the tunnel. We saw that both pulses simultaneously hit the tunnel, and we denote by B the point of the tunnel that receives both pulses.

Q1: Represent this scenario with a Minkowski diagram. We call O* the common origin-event: "you and your friend emit a pulse". Your train frame is K', whereas the ground is K. We denote by B* the event "both light pulses reach the tunnel at the point B".

Q2A: Show the distance covered by the pulse, seen from the ground.

Q2B: Show the distance covered by the pulse, seen from the train.

Q3A: Show on the diagram the proper distance between the point O and the entrance of the tunnel.

Q3B: Is the distance train-tunnel at the time t' = 0 equal to the distance covered by the pulse, seen from the train?

Q4: Represent in the Minkowski diagram the distance train-tunnel seen from the train at t' = 0.

Q5: How can you explain the difference between the distance train-tunnel seen from the train at the time t' = 0, and the distance covered by the pulse emitted from the train (cf. Question 2)?

Notes

1 These statements which seem natural are demonstrated in Volume II Section 1.4.2.1.
2 This statement also seems natural, and is demonstrated in Volume II Section 1.2.2.
3 This experiment is described in Volume II Section 1.5.1.
4 Cf. more in Section 7.1.

3

Other New Kinematics Laws, Causality and Accelerated Trajectories

Introduction

Having seen that the Galilean coordinate transformation law was wrong and replaced by the Lorentz transformation, the classical velocity addition law must also be wrong. Hence, the new velocity composition law will be demonstrated and its important consequences will be examined. In particular, we will see that c is the only invariant speed, and that it cannot be exceeded without contradicting the fundamental principle of causality. Next we will show that our space-time universe possesses a metric which is different from the classical one, and has amazing properties. We will then examine the case of accelerated objects, with in particular the famous Langevin's twin scenario.

3.1 The New Velocity Composition Law and Consequences

We will first establish the new velocity composition law from the Lorentz transformation. We will then see its properties and amazing consequences.

3.1.1 The New Velocity Composition Law

Consider an object which is moving in any direction; its trajectory is a succession of events; we will consider two close events, M* and (M+dM)*, defining the vector $\overrightarrow{dM^*} = [M^*, (M+dM)^*]$ (Figure 3.1).

This same vector $\overrightarrow{dM^*}$ is seen in K as $\overline{dM}$ (dx, dy, dz, dt), and in K' as $\overline{dM'}$ (dx', dy', dz', dt'). The velocity of this object in K' is denoted by $\overline{W'}$, with its coordinates along O'X' and O'Y' denoted by W'_x and W'_y. We thus have: $W'_x = dx'/dt'$ and: $W'_y = dy'/dt'$.

Similarly, the velocity of this object in K is denoted by $\overline{W}$, with its coordinates along OX and OY being W_x and W_y. We have: $W_x = dx/dt$ and: $W_y = dy/dt$.

DOI: 10.1201/9781003201335-3

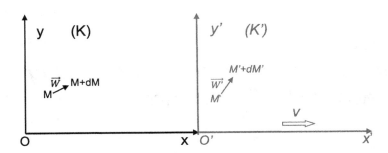

FIGURE 3.1
The object at the event M* has the velocity $\vec{w}$ in K and $\overrightarrow{w'}$ in K'.

The velocity $\vec{W}$ is the result of composition of the object velocity $\overline{W'}$ with the velocity $\vec{V}$ of the frame K' relative to K. To calculate $\vec{W}$ knowing $\overline{W'}$, we will calculate the coordinates of $\overline{dM}$ knowing those of $\overline{dM'}$: we saw that this is obtained by applying the reverse Lorentz transformation: $\overline{dM} = (\Lambda_{-v})d\overline{M'}$, which is:

$$dx = \gamma\left(dx' + vdt'\right) = \gamma dt'(W'_x + v)$$

$$dt = \gamma\left(dt' + dx'.v/c^2\right) = \gamma dt'(1 + v.W'_x/c^2), \text{ and: } dy = dy'.$$

We can then calculate the components W_x and W_y of the object velocity $\vec{W}$ seen in K:

$$W_x = \frac{dx}{dt} = \frac{\gamma dt'\left(W'_x + v\right)}{\gamma dt'\left(1 + v W'_x/c^2\right)}, \text{ so finally: } \mathbf{W_x = \frac{W'_x + v}{1 + vW'_x/c^2}} \qquad \blacksquare (3.1)$$

$$W_y = \frac{dy}{dt} = \frac{dy'}{\gamma \, dt'\left(1 + v W'_x/c^2\right)}, \text{ so finally: } \mathbf{W_y = \frac{W'_y}{\gamma\left(1 + vW'_x/c^2\right)}} \qquad \blacksquare (3.2)$$

We can see that the velocity composition law is no longer additive like the classical Galilean law, which was: $\vec{W} = \vec{V} + \overline{W'}$. Hence, the wording velocity addition is no more appropriate, but velocity composition is preferable. (Velocity *combination* is also used.) Similarly, the symbol $\oplus$ appears more suitable for velocity composition since it merges the composition and the addition symbols. Hence, we will write: $\vec{W} = \vec{V} \oplus \overline{W'}$.

We first notice that for usual speeds far from c, this new law is very close to the classical Galilean law.

We will first examine the relation 3.2, and then the relation 3.1 which has very important consequences.

3.1.1.1 The Transverse Components of the Object Velocity Are Not Invariant

In the case where W'_x is null, meaning that the object moves along the O'Y' direction, the relation 3.2 gives: $W_y = W'_y / \gamma_v$. Unlike transverse distances, the transverse velocities are not invariant, which is a novelty.

Moreover, in the general case where W'_x is not null, the transverse component W_y depends on W'_x. This implies that one cannot *separately* consider the components of the velocity W' when making the velocity composition, and the new velocity composition law is therefore not linear.

We will now focus on equation 3.1 and its important consequences:

3.1.2 Composition of Speeds Parallel to the Speed of the Frame K' Relative to K

The equation 3.1 gives, as a particular case, the composition law for speeds that are parallel to the speed v of the frame K'. For the sake of simplicity, we will assume the object moves along a parallel line to the O'X' = OX axis; we will denote by w' its speed in K', and w in K. The equation 3.1 thus becomes:

$$\mathbf{w} = \mathbf{v} \oplus \mathbf{w'} = \frac{w' + v}{1 + w'.v / c^2}. \tag{3.3}$$

3.1.2.1 The Velocity Composition Curve

The function which gives w knowing w' according to equation 3.3 is a homographic function, denoted by: $w = f_v(w')$. The curve below represents the homographic function:

$$w = f_v(w') = v \oplus w' = \frac{w' + v}{1 + w'.v / c^2}.$$

We know that for w'=0, we have: w=v, and that for w'=−v, we must have: w=0.

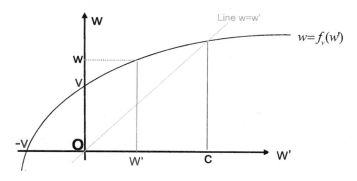

FIGURE 3.2
Velocity composition curve.

We also know that for usual speeds, the function must be very close to: $w=v+w'$; hence, the tangent for $w'=0$ must be very close to 1. (It is the derivative of $f(w')$ at $w'=0$, which is: $1-v^2/c^2$.)

3.1.2.2 The Velocity's Composition Law is Commutative

We notice that v and w' play identical roles; hence: $w=v \oplus w'=w' \oplus v$. This also means that the velocity composition law is commutative.

As an illustration, if an object is going at the speed w' inside a train (frame K') that is going at the speed v relative to the ground (frame K), then this object is going at the same speed relative to the ground as another object which goes at the speed v inside a train that is going at the speed w'.

3.1.2.3 Case of Negative Speed v

The equation 3.1 is valid whatever the sign of W'_x, and likewise w'. If the sign of w' is negative, let's call w'' its absolute value, and we have: $w''=-w'$.

We want to calculate $w=v \oplus w'=v \oplus (-w'')$. Let's use the commutativity of the equation 3.1: we are in the case where the frame K' goes at the speed $-w''$, whereas the object goes at the speed v; hence, we have:

$$w=(-w'') \oplus v=(v-w'')/(1-w''v/c^2).$$

3.1.2.4 The Reverse Velocity Transformation Law

We want to know the speed w' in K' of an object which is moving' at the speed w in K: we can consider that w' is the composition of the speed w with the speed of K relative to K', which is $-v$, and we thus have:

$w'=(-v) \oplus w=(w-v)/(1-wv/c^2)$. This is again a homographic function: $W=f_{-v}(w')$.

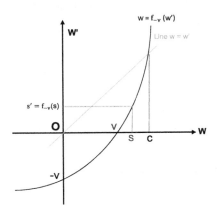

FIGURE 3.3
Reverse velocity composition curve.

We will now see very important consequences of this velocity composition law: we will first show that c is the only invariant speed, and that it cannot be exceeded by combining speeds below c.

3.1.3 The Speed c Is Both the Only Invariant Speed and the Maximum One

3.1.3.1 *The Speed c Is the Only Invariant Speed*

The curve of the homographic function $w = f_v(w')$ has only one intersection with the line $w = w'$, as shown in Figure 3.2. Since we know that the speed c is invariant, c must be this single intersection (Figure 3.3).

> This can be mathematically demonstrated: if another speed were invariant, say d this invariant speed, and we could write: $d = (d+v)/(1+dv/c^2)$; then: $d.(1+dv/c^2) = d+v$; hence: $d^2v/c^2 = v$. Consequently, $d = \pm c$, meaning that c and −c are the only possible invariant speeds.

3.1.3.2 *The Speed c cannot be Exceeded by Composing Speeds Below C*

We can think of surpassing the speed c by launching a very fast rocket going at the speed of 0.5 c, and from this rocket, we launch a very fast projectile at the speed of 0.7 c in the same direction as the rocket (for example, fast accelerated electrons). The velocity composition curve (Figure 3.2) shows that the speed of the electrons relative to the ground will be lower than c. The calculation confirms:

$$w = (0.5c + 0.7c)/(1 + 0.5 \times 0.7) = 1.2c/1.35 = 0.889c .$$

Mathematically, it can be shown that this result is general: there cannot exist a speed w' which is below c and such that: $w = f_v(w) > c$. Indeed, the function f_v is continuous and monotonically increasing; hence, any speed w' which is below c will have its image below $f_v(c) = c$.

The fact that f_v is monotonically increasing can be seen by the sign of its derivative, as we indeed have: $f'_v(w') = [(1+vw'/c^2) - v/c^2(v+w')]/(1+w'v/c^2)^2$. The numerator, $1 - v^2/c^2$, is always positive since $v < c$. (We recall that v is the speed of K' relative to K.) The denominator is always positive too.

3.1.3.3 *Further Arguments Showing That C Is the Maximum Speed*

First, no greater speed than c has ever been observed (from an inertial frame[1]). Besides, there are further theoretical arguments preventing the possibility of greater speeds than c, in particular:

- We will see in the chapter on the energy (Section 5) that the energy required for any object to reach c is infinite, except for massless objects such as the photon.

- We will see in the next chapter that an object going at a greater speed than c, called Tachyon, is seen in another inertial frame arriving before departing, which is absurd and contradicts the fundamental causality principle. Furthermore, Tachyons cannot have a real existence, as shown in Volume II Section 7.3.2.

3.2 The Causality Principle and the Three Categories of Space-Time Intervals

3.2.1 The Causality Principle and the Relativistic Causality Condition

The causality principle states: **a consequence cannot precede its cause.** It is an essential property of our universe; however, it cannot be demonstrated and therefore it is a postulate. Causality also means that time always goes in the same direction, we cannot reverse time, and we cannot intervene in the past. Noteworthy, the causality principle was never used in all previous demonstrations of Relativity, including the Lorentz transformation; hence, one may wonder if Relativity respects this fundamental postulate.

Consider a simple example of causality: You press a switch that closes an electrical circuit, and as a result, a lamp lights up. The change in the lamp must occur *after* the button was pressed.

However, Relativity showed that the notions of past, present and future are relative to an inertial frame. Hence, one may wonder if there can exist a frame where the lamp lights up *before* the button was pressed. To show that this issue is real, consider the following scenario:

3.2.1.1 Scenario Showing That a Greater Speed than c Contradicts the Causality Principle

The switch is at the point A, and the lamp at the point B, both in an inertial frame K. We turn on the switch at the time t=0 of K, and at that time, an object M is going from A to B at the constant speed S. Then, when M arrives at B, it immediately triggers the lamp ignition.

Let's have another frame K′ going at the speed v relative to K in the same direction as the segment AB. We set the common origin-event of K and K′ to be: "the object M is in front of the point A"; and the segment AB is along the OX and O′X′ axes (Figure 3.4).

In K, M is in front of B at the time $T_b = AB/S$. In K′, it is at the time T_B', which is given by the Lorentz transformation: $T_B' = \gamma(T_b - AB.S^2/c) = \gamma(AB/S - AB.S^2/c)$.

We notice that T_B' is not always positive. Indeed, if S>c, then AB/S<AB/c and $AB.S^2/c > AB/c$. Consequently, $AB/S - AB.S^2/c < AB/c - AB/c = 0$; hence: $T'b < 0$, which means that in K′, M is seen arriving in B before being in A.

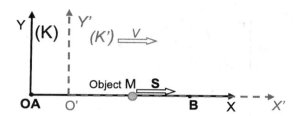

FIGURE 3.4
The object M seen from K and K'.

The lamp is on before the switch was pushed! This contradicts the causality principle, and for such possibility to be avoided, no object should be able to go at a faster speed than c.

More generally, if an event E* is the cause of an event F*, there must be a minimum delay between these events: in any frame K, E* and F* are seen in the points denoted by E and F; then, this delay in K corresponds to the time taken by the fastest object, i.e., the photon, to go from point E to point F. This will be referred to as the *relativistic causality condition*.

Still, we need to show that if the events E* and F* respect this relativistic causality condition in one inertial frame, K, then they respect this condition in all inertial frames. In other words, we need to show that this causality condition is intrinsic. To this end, we will first present a demonstration based on the Minkowski diagram, and then a mathematical one.

3.2.2 Causality Is an Intrinsic Notion, and the Three Types of Space-Time Intervals

We will show with the Minkowski diagram that if this relativistic causality condition is respected in one frame, K, then it is respected in all frames: let's choose E* as the origin-event of the frame K. We thus have: $X_e = T_e = 0$. The time taken by a photon to go from O to F* is: X_f/c. Then for this photon to arrive before the event F*, or simultaneously with F*, we must have: $T_f \geq X_f/c$; the relativistic causality condition then reads:

$$cTf \geq Xf \qquad\qquad ■ (3.4)$$

In the Minkowski diagram, this inequality means that the point F which represents the event F* having coordinates in K (X_f, cT_f) is located in the upper part above the line of light.

Let's now see these events E* and F* from another frame K' which shares E* as common origin-event: K is moving at the speed v relative to K, which implies that the angles of the axis of the Minkowski diagram verify: $\alpha = (OX, O'X') = (O'cT', OcT)$ with $tg\alpha = v/c$. The event F* is seen at the same point F* in both K and K', meaning that this point F* is located above the line of light of K' since it is the same line of light as for K.[2]

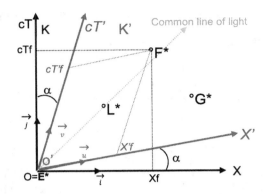

FIGURE 3.5
The event F* is above the line of light of all inertial frames. Similarly, the event G* is below the line of light of all inertial frames.

Consequently, we have cT'f>X'f, which means that the couple (E*, F*) verifies the relativistic causality condition in K'.

Causality thus is an intrinsic notion provided that the relativistic causality condition is respected.

Any vector E*F* which respects this relativistic causality condition is called **"time-like"** because the point F* is closer to the time axis (OcT) than to the distance axis (OX). This property is intrinsic, meaning that if a vector is time-like in one inertial frame, it is also time-like in all inertial frames.

In the above switch and lamp scenario, the events E* and F* form a time-like vector because the action of pressing a switch triggers the motion of electrons in an electrical circuit from the switch to the lamp, and as these electrons move at a speed which is lower than the speed of light, the relativistic causality condition is respected.

Conversely, all points which are below the line of light, like the event G* in Figure 3.5 above, correspond to events which do not verify the relativistic causality condition, since cTg<Xg in all inertial frames. Hence, there cannot be a causality relationship between the events E* and G*. Such vectors E*G* are called **"space-like"** because the point G is closer to the distance axis (OX) than to the time axis (OcT).

Finally, points which are on the line of light (yellow), like the event L* in Figure 3.5 above, represent events belonging to the trajectory of a photon (light): cT=X; these vectors E*L* are called **"light-like"**.

Figure 3.6 below shows the possible positions of future and past events relative to an observer, using the same structure as the Minkowski diagram, but adding one dimension along the OY axis.

The future events which can be caused (or influenced) by an observer are inside the upper light cone whose center is the observer. The frontier of this cone is made of the lines of light: $x^2+y^2=c^2t^2$.

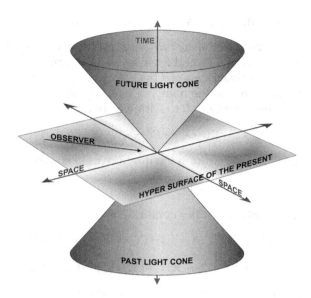

FIGURE 3.6
The light cone. (Image from Shutterstock.)

The events that are in the past of the observer are inside the lower light cone (with the caption "Past Light Cone" in Figure 3.6 above). These events could have had an influence on the observer.

All other events that are neither in the past nor in the future of the observer have no causal relationship with him. In some frames they are in his future, in some others in his past, and in some frames they are even simultaneous.

Remark 3.1: Relativity thus shows the impossibility of instantaneous distant interactions. However, at the time Special Relativity was discovered, gravitation was thought to exert instantaneous interactions, even over huge distances across the universe. This was a serious objection to the universality of the theory of Relativity, which induced Einstein to show with his General Relativity theory that gravitation makes no exception to Relativity: he indeed showed that gravity propagates at light speed (like electromagnetism).

Remark 3.2: Causality in common language has several meanings, including the notion of responsibility, which in turn refers to the issue of determinism. Relativity is agnostic regarding these issues: the event E* is the cause of the event F* if E* is part of a process that leads to the existence of F* (cf. more in Volume II Section 7.1.2).

Remark 3.3: The reader may deduce that for an event E* to be the cause of an event F*, there must exist an object (electron, photon, etc.) that physically goes from E* to F*, or a perturbation of a physical medium which propagates from E* to F*. Relativity does not explicitly imply this, but it is *as if* such a physical entity was indeed needed.

3.2.3 Causality and the Space-Time Interval Invariance

Let's translate the above relativistic causality condition into relations between the coordinates of the events E* and F*: in an inertial frame K, the coordinates of E* and F* are, respectively, denoted by: $E^*(X_e, Y_e, Z_e, T_e)$ and $F^*(X_f, Y_f, Z_f, T_f)$. These events are seen in K at the points $E(X_e, Y_e, Z_e)$ and $F(X_f, Y_f, Z_f)$. The time ΔT taken by a photon to go from E to F is given by:

$$c^2 \Delta T^2 = (X_f - X_e)^2 + (Y_f - Y_e)^2 + (Z_f - Z_e)^2.$$

The relativistic causality condition in equation 3.4 becomes:

$$c^2 (T_f - T_e)^2 > \Delta T^2.$$

So finally, the relativistic causality condition is:

$$c^2 (T_f - T_e)^2 > (X_f - X_e)^2 + (Y_f - Y_e)^2 + (Z_f - Z_e)^2.$$

$$\text{Or: } c^2 (T_f - T_e)^2 - (X_f - X_e)^2 - (Y_f - Y_e)^2 - (Z_f - Z_e)^2 > 0. \tag{3.5}$$

We will present a second demonstration showing that if this condition is respected in one frame K, it is respected in all inertial frames. We will first demonstrate an important property: the invariance of the Lorentzian norm of any space-time separation vector between two events E*F*.

The expression $c^2 (T_f - T_e)^2 - [(X_f - X_e)^2 + (Y_f - Y_e)^2 + (Z_f - Z_e)^2]$ is called the **space-time interval**, denoted by: $ds^2(E^*F^*)$ or $\Delta s^2(E^* F^*)$ or even $(E^*F^*)^2$ when there is no risk of confusion with the classical norm. This expression is also the square of the **Lorentzian norm**. We thus have:

$$ds^2 (E^*F^*) = c^2 (T_f - T_e)^2 - \left[(X_f - X_e)^2 + (Y_f - Y_e)^2 + (Z_f - Z_e)^2 \right] \quad \blacksquare (3.6)$$

We can see from equation 3.5 and equation 3.6 that the sign of $ds^2(E^*F^*)$ determines the possibility of a causal relationship between E* and F*:

- If $ds^2(E^*F^*)$ is positive, there is a possibility of causation between E* and F*, and E*F* constitutes a time-like interval.
- If $ds^2(E^*F^*)$ is negative, there is no causation possibility, and E*F* constitutes a space-like interval.
- If $ds^2(E^*F^*)$ is null, E*F* constitutes a light-like interval. All such events F* form a cone of light, as shown in Figure 3.6, which separates the universe between the region where causation with the event E* is possible and the region where it is not.

We will now show that the expression in equation 3.6 gives the same result in all inertial frames, meaning that the space-time interval is invariant.

This will imply that the sign of $ds^2(E*F*)$ is the same in all inertial frames, meaning that the relativistic causality condition is intrinsic.

3.2.3.1 The Lorentzian Norm Invariance: Mathematical Demonstration

Let's express $ds^2(E*F*)$ in another frame K' moving at the velocity v relative to K. The origin-event being arbitrary, we choose E* as the common origin-event of K and K', with the axes OX and O'X' parallel to the velocity v. The coordinates of E* in K and in K' are: (0, 0, 0, 0). The coordinates of F* are denoted by (X, Y, Z, T) in K and (X', Y', Z', T') in K'. The space-time interval E*F* is then in K: $ds^2(E*F*) = c^2T^2 - X^2 - Y^2 - Z^2$.

In K', the coordinates of F* are given by the Lorentz transformation:

$$X' = \gamma (X - vT); \; Y' = Y; \; Z' = Z; \; T' = \gamma (T - v/c^2 X).$$

The space-time interval E*F* is in K':

$ds'^2(E*F*) = c^2\gamma^2(T - v/c^2 X)^2 - \gamma^2(X - vT)^2 - Y^2 - Z^2$.

Then: $ds'^2(E*F*) = c^2\gamma^2 T^2 - 2\gamma^2 vXT + \gamma^2\beta^2 X^2 - \gamma^2 X^2 - \gamma^2 v^2 T^2 + 2\gamma^2 vXT - Y^2 - Z^2$.

So: $ds'^2(E*F*) = c^2\gamma^2 T^2(1 - \beta^2) + X^2\gamma^2(\beta^2 - 1) - Y^2 - Z^2$. Then, having: $\gamma^2(1 - \beta^2) = 1$, we finally obtain: $ds'^2(E*F*) = c^2 T^2 - X^2 - Y^2 - Z^2 = ds^2(E*F*)$ ∎

3.2.4 Properties of Time-Like and Space-Like Intervals

3.2.4.1 Properties of Time-Like Intervals

3.2.4.1.1 Existence of an Inertial Frame Where the Two Events Are Co-Located

We will show that for any time-like interval E*F*, there is one inertial frame where these two events are co-located. (Another demonstration is proposed in Problem 3.2 using Minkowski diagram.)

In an inertial frame K, these events are seen at the points E and F, and at the times T_E and T_F. We set: $T_{EF} = T_F - T_E$ and we know that $T_{EF} > 0$ since E*F* is a time-like interval.

Let's have an inertial object M which is seen in K passing by the points E at the time T_E and F at the time T_F. The speed of this object is then: $v = EF/T_{EF}$. This speed v is below c because E*F* is a time-like interval, meaning that: $c.T_{EF} > EF$, and then: $v = EF/T_{EF} < c$.

For this inertial object M, the two events E* and F* are co-located: they are both located at object M but at different times in its inertial frame. That is, these two events are co-located in the inertial frame Ko where this object M is fixed, meaning that Ko is moving at the speed $v = EF/T_{EF}$ relative to K.

This will enable us to show other important properties of time-like intervals.

3.2.4.1.2 *The Space-Time Interval Represents the Duration of an Inertial*
Journey from E to F*, as Measured by the Object's Proper Time*

The time elapsed between E* and F* and measured by the object M is a proper time of this object, denoted by: $(\tau_F - \tau_E) = \tau_{EF}$. It is also the time of the frame Ko since this object is fixed in Ko.

We can calculate $ds^2(E^*F^*)$ in any inertial frame knowing that the Lorentzian norm is invariant. Let's then calculate it in Ko, and we obtain:

$$ds^2(E^*F^*) = c^2.\tau_{EF}^{\,2} \quad \blacksquare$$

3.2.4.1.3 *The Space-Time Interval Is the Shortest Time between*
E and F* among All Inertial Frames*

Consider any couple of events (E*, F*). In any inertial frame K, these events are seen, respectively, in the points A and B. The time duration in K between these two events is denoted by T_{EF}. We have: $ds^2(E^*F^*) = c^2.T_{EF}^2 - AB^2$; on the other hand, we saw that: $ds^2(E^*F^*) = c^2.T_{EF}^2$. Hence, T_{EF}^2 is always greater than τ_{EF}^2, except in Ko, and this is consistent with the time dilatation law.

We will now show the corresponding properties of space-like intervals.

3.2.4.2 **Properties of Space-Like Intervals**

We will first show that for any space-like interval (E*, F*), there exists a frame where these two events are simultaneous, and we will then draw some important consequences:

3.2.4.2.1 *Existence of an Inertial Frame Where These Two*
Events E and F* Are Simultaneous*

Consider a frame K where the origin-event is E*, and F* is on the OX axis. The coordinates of F* in K are: F(Xf, Tf). In an inertial frame K' that is moving at the velocity v relative to K, and also having E* as origin-event, the coordinates of F* are given by the Lorentz transformation. We thus have for its time coordinate T'f:

   $T'f = \gamma.(Tf - v/c^2.Xf)$. Then if we choose $v = c^2Tf/Xf$, we have: $T'f = 0 = T'e = Te$. We denote by Ko this frame which is going at the speed $v = c^2Tf/Xf$ relative to K along its OX axis. This result shows that in Ko, the events E* and F* are simultaneous.

We can check that: $v < c$. Indeed, E*F* being a space-like interval, we have: $Xf > cTf$, and then:

$$v = c^2Tf/Xf < c^2Tf/cTf = c$$

3.2.4.2.2 *The Space-Time Interval $ds^2(E^*F^*)$ Is the Negative of the*
Classical Distance² between E and F in the Frame Ko

In the frame Ko, we denote by E and F the points where E* and F* are seen. We have:

$$ds^2(E^*F^*) = c^2(Tf - Te)^2 - EF^2 = -EF^2$$

The distance EF is a proper distance since it is measured in the frame Ko where E* and F* are simultaneous. In other words, ds²(E*F*) is the negative of the square of the distance EF measured in the frame Ko where E* and F* are simultaneous.

3.2.4.2.3 EF Is the Shortest Classical Distance between
E* and F* among All Frames

In any other inertial frame K than Ko, let's call P and Q the points where the events E* and F* take place. We have: $ds^2(E^*F^*) = c^2 T_{ef}^2 - PQ^2$, and we saw that: $ds^2(E^*F^*) = -EF^2$, hence:

$$PQ^2 = EF^2 + c^2 T_{ef}^2.$$

Remark 3.4: This does not contradict the length contraction law since PQ is not the length EF seen from K, because E* and F* are not simultaneous in K.

3.3 The New Metric in the Space-Time Universe

The Lorentzian norm invariance is a fundamental result of Relativity: it is the new metric of our space-time universe. We will first show what this new metric geometrically represents with another demonstration based on the famous time dilatation scenario. We will then see its properties, including the amazing triangle inequality which implies a significant change of the fundamental law of inertia. The importance of this novelty will be even greater in General Relativity.

3.3.1 The Lorentzian Norm Invariance: Physical Demonstration

The following demonstration has the interest of being physically oriented and of reusing the famous time dilatation scenario (cf. Section 2.1.2). It also shows that the Lorentzian norm invariance is a consequence of the light speed invariance postulate.

The same perfect clock is made in the point O' of the inertial frame K'. A mirror is placed in the point M', with O'M' being perpendicular to the O'X' axis of K'.

Let's have a first event E: "a light pulse is emitted from O' toward M'". Then, a second event F: "this light pulse has returned to O' after having been reflected by the mirror in M'".

We will show that the Lorentzian norm² of EF is the same in K' and in another inertial frame K moving relative to K' along the O'X' axis.

In K', these two events are co-located; hence, the time between them measured is a proper time, denoted by τ_{EF}. The Lorentzian norm² of EF in K' is:
$ds_{K'}^2(EF) = c^2 . \tau_{EF}^2 - O'O'^2 = c^2 . \tau_{EF}^2.$

Let's have a geometrical picture of this norm: the length of the segment O'M' in K' is: $\frac{1}{2}.c.\tau_{EF}$ (since the pulse has covered twice the distance O'M' during τ_{EF}). We thus have: $ds_{K'}^2(EF)=4.O'M'^2($, *and we recall that O'M' is the proper length between the mirrors* (Figure 3.7).

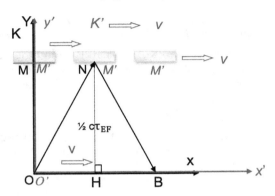

FIGURE 3.7
Moving perfect clock.

Let's now consider this scenario from the frame K: the frame K' is seen moving relative to K at the speed V along the OX of K (and the O'X' axis of K'). In K, the event E occurs in O at t=0, and the event F occurs at the point B at the time T_f. The Lorentzian norm² of EF in K is by definition:

$$ds_K^2(EF)=c^2T_f^2 - OB^2.$$

The pulse has covered the segment ON and then NB (=ON) during the time T_f; hence, we have: $T_f=2.ON/c$; subsequently:

$$ds_K^2(EF)=4\ ON^2-OB^2=4(ON^2-OH^2).$$

In the rectangle triangle OHN, the Pythagorean relation gives: $ON^2=OH^2+HN^2$; hence: $ON^2-OH^2=HN^2$.

Then finally, $ds_K^2(EF)=4.HN^2=4.O'M'^2=ds_{K'}^2(EF)=c^2.\tau_{EF}^2$.

This result gives a geometrical meaning to the Lorentzian norm:

$$\sqrt[2]{ds^2(EF)}=2.O'M'=c.\tau_{EF} \ \blacksquare$$

We can thus see that the Lorentzian norm is equal to twice the proper length of the segment of Einstein's perfect clock. Furthermore, this demonstration can be generalized to any time-like interval, as we saw that for any couple of events (A*, B*), there exists a frame Ko where they are co-located (cf. Section 3.2.4.1.1).

Remark 3.5: Given the importance of the Lorentzian norm, other demonstrations of its invariance are presented in Volume II, revealing different aspects: a mathematical one in Section 2.1.1; one based on Minkowski's complex space in Section 2.7.5; and one based on the Lorentz transformation being a hyperbolic rotation in Section 2.5.2.

3.3.2 The New Metric

In classical physics, the metric of the 3D space was the length of a segment, and it is a practical necessity. Mathematically, a metric is a function that associates a real number with any couple of points (vector), and this number is independent of the frame where this vector is considered. The metric is thus an intrinsic notion. In classical physics, the length of a segment was the same independently of the frame where it is considered, but Relativity showed that this is no longer the case (cf. the length contraction); hence, the classical distance definition can no longer constitute a metric.

Minkowski showed that we are in a 4D space-time universe, and that the elementary notion is the event. He then showed that the new metric between any couple of events (E*, F*) is the space-time interval since it is the only[3] function that associates a number to a couple of events in an intrinsic way (i.e., giving the same number in all inertial frames).

We saw that the sign of the Lorentzian norm[2] discriminates between the three categories of space-time intervals:

- If $ds^2(E*F*)>0$, then E*F* is a time-like interval; there is a possible causal relation between E* and F*; and $\sqrt[2]{ds^2\left(E^*F^*\right)}$ measures the proper time duration taken by an inertial object to go from E* to F*.
- If $ds^2(E*F*)<0$, then E*F* is a space-like interval; there is no causal relation between E* and F*; and $\sqrt[2]{-ds^2\left(E^*F^*\right)}$ measures the length EF in the frame where E* and F* are simultaneous (Ko).
- If $ds^2(E*F*)=0$, then E*F* is a light-like interval; there can be a photon going from E* to F*.

All events that are equidistant to an event O* do not form a sphere any longer, but a 4D hyperbolic surface, since any event M* such that $ds^2(O*M*)=cst$ satisfies the relation: $c^2t^2-(x^2+y^2+z^2)=cst$, which is the equation of a hyperbolic surface, as shown below for the case where $cst>0$ (Figure 3.8).

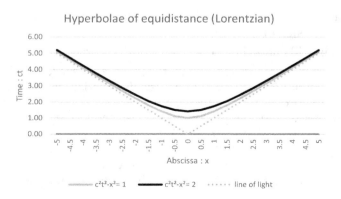

FIGURE 3.8
The events that are equidistant to a given event are on a hyperbolic surface.

The new metric has surprising properties, in particular the triangle inequality which is the opposite of the classical one: take three random events: A*, B* and C*. In classical physics, the distance between AB is always shorter than the sum of the distance AC and the distance CB, and this reflects the fact that the straight line is the shortest distance between two points. With the Lorentzian distance, it is the opposite: $\sqrt[2]{ds^2(A^*B^*)} > \sqrt[2]{ds^2(A^*C^*)} + \sqrt[2]{ds^2(C^*B^*)}$. This property will be demonstrated in Section 3.5.5, and its important consequence will be presented. In particular, the formulation of the fundamental law of inertia will change. However, it is in General Relativity that the new metric will be most important, because the chrono-geometry of our universe will be distorted; hence, it will lose its Euclidean characteristics, particularly the straight line, but it will keep the Lorentzian distance which will be a major property.

Remark 3.6: Notation considerations: the notation $ds^2()$ refers to small or even infinitesimal quantities, as is the case in General Relativity. In Special Relativity, conversely, the laws also apply on long distances; hence, the notation $\Delta s^2()$ is also used to express space-time intervals that are not infinitesimal.

3.4 The Time-Distance Equivalence

Looking at the previous chapter, we notice that time and distance play symmetrical roles in several major laws:

- The Lorentzian norm invariance law: consider a space-time separation vector: $MN = (\Delta X, \Delta T)$. We notice that the term $\Delta(cT)$ plays a parallel role to ΔX in the law:

$$\Delta s^2(MN) = c^2(\Delta T)^2 - (\Delta X)^2 = c^2(\Delta T')^2 - (\Delta X')^2.$$

- What discriminates between a time-like interval and a space-like one is the sign of its norm^2.
- The time dilatation law versus the length contraction.
- The symmetrical properties of time-like and space-like intervals (cf. Section 3.2.4). In particular, the Lorentzian norm represents either the proper time in the frame where the events are co-located, or the negative of the proper distance in the frame where they are simultaneous.

These similarities and symmetrical roles of time and distance mean that there is a form of equivalence between time and distance. Moreover, this equivalence also appears with another surprising property that we will see now: the possibility of having a single unit for time and distance.

3.4.1 Time and Distance Can Be Naturally Expressed with a Single Unit

The invariance of Lorentzian norm^2, $\Delta s^2() = (\Delta cT)^2 - (\Delta X)^2$, invites us to consider the product cT as a time expressed with a distance unit because for any sum to make sense, each term must have the same dimension and the same unit.

Within any frame K, if we want to express a time duration t with a distance unit, we must multiply the time with a speed. If we want this method to be universal, meaning the same in any frame while giving the same result, it must use a speed that is the same in all frames. Hence, c is the natural choice determined by the space-time geometry.

This universal time to distance conversion method, $t \rightarrow ct$, is not only a change of unit (like passing from km to miles), but also a change of dimension. We can even say that using a single unit for time and distance is the *natural* way of expressing these quantities: indeed, since "Mother Nature" enables us to use only one unit, using two brings some complexity, and we will indeed see that physical expressions are simpler with the single unit for time and distance. Using a single unit is often referred to as "natural" or "homogeneous" units. This novelty of adding time and distance quantities will be further encountered in other physical laws.

Note that the common unit can also be a time unit. For instance, the light-year is commonly used in astronomy. We could simply say that the unit of distance is the year, but the wording light-year facilitates realizing the magnitude of the distances. What really matters in a natural system of units is that the photon follows the line: $T=X$.

Note that the word equivalence can be misleading:

3.4.1.1 Equivalence Does Not Mean Identity

The name "equivalence" may sound curious to some readers since time and distance remain fundamentally different: in particular, time still has only one dimension with an orientation (cf. the causality principle), and incurs the time dilatation phenomenon. Conversely, distance still has three dimensions without any orientation, and incurs the length contraction. Even if time and distance share a common unit, time-like intervals have the opposite sign as space-like intervals.

Mathematically, equivalence is different from identity. Mathematically, a relation of equivalence is reflexive, symmetric and transitive. It is different from identity. In physics, equivalence is often used when two quantities are always proportional, which is the case when writing: $d = ct$.

3.4.2 Physical Laws Are Simpler with Natural Units

We will show with two important examples that using natural units brings more simplicity. However, we will first indicate the guidelines for converting expressions into the natural units system.

3.4.2.1 *The Conversion Method $t \to ct$*

Let's start with the time t expressed in second, and the distance d in meter. The conversion for the time consists of multiplying the time t by c expressed in m/s (i.e., $\approx 3 \times 10^8$ m/s):

$t \to t_d = t.c$ (t_d means the time expressed in natural units, which is distance in this case).

The distance remains identical since we have chosen to express the time with the distance unit (the meter).

A speed is a distance divided by a time. In natural units, both share the same unit; hence, a speed now has no dimension. The speed v of any object becomes in natural units: $v_n = v/c = \beta$ (v_n means the speed v expressed in natural units, but it is generally denoted by β).

Indeed, the speed v of any object is the distance in meters covered during 1 second. This duration of 1 second expressed with natural units is: 1.c. Hence, the speed expressed in natural units is: $v_n = v/1c = \beta$.

In particular, the speed of light c becomes $c_n = 1$ without dimension (c_n means the speed of light in natural units, but it is generally denoted by c when there is no risk of confusion).

One consequence is that the gamma factor remains identical since it involves $v/c = \beta$, which is the ratio between two speeds v and c. (It already had no dimension in classical units.)

3.4.2.2 *The Lorentzian Norm Expressed with Natural Units, and its Invariance*

To avoid confusion, we will denote by t_d the time t expressed with natural unit: we thus have: $t_d = c.t$. The ds^2 invariance law becomes:

$$ds^2() = t_d^2 - (x^2 + y^2 + z^2) = t_d'^2 - (x'^2 + y'^2 + z'^2).$$

With the following notations for the 3D distances, $d = x^2 + y^2 + z^2$ and $d' = x'^2 + y'^2 + z'^2$, this law reads:

$$ds^2() = t_d^2 - d^2 = t_d'^2 - d'^2.$$

We notice that the $ds^2()$ invariance law is simpler, and that time and distance play symmetrical roles.

3.4.2.3 The Lorentz Transformation Expressed with Natural Units

To express the Lorentz transformation with natural units, we will use the rules indicated in Section 3.4.2.1:

The relation: $x' = \gamma.(x - v.t)$ becomes: $x' = \gamma.(x - v/c.ct)$ or: $\mathbf{x'} = \boldsymbol{\gamma.(x - \beta.\,t_d)}$ ∎

The relation: $t' = \gamma.(t - x.v/c^2)$ becomes: $ct' = \gamma.(ct - x.v/c)$ or: $\mathbf{t'_d = \gamma.(t_d - \beta.x)}$ ∎

So finally:

$$\begin{pmatrix} x' \\ t'_d \end{pmatrix} = \gamma \begin{pmatrix} 1 & -\beta \\ -\beta & 1 \end{pmatrix} \begin{pmatrix} x \\ t_d \end{pmatrix} \tag{3.7}$$

where $t_d = c.t$ and $t'_d = c.t'$.

We notice the simplicity of the Lorentz transformation with natural units. It also reflects the symmetrical roles of time and distance, as previously seen.

More generally, the simplicity of expressions with natural units stems from the fact that they do not carry the complexity due to introduction of an unnecessary unit that distinguishes time from distance.

3.4.2.4 Some Reservations as to the General Use of Natural Units

In our day-to-day activities, using a single unit for time and distance would be extremely impractical: it would force us to manipulate either extremely huge numbers (with a common distance unit, 1 second becomes 1.c) or extremely small ones. However, when considering travels in the cosmos, the single unit "light-year" for both the time and the distance is very practical and commonly used.

Moreover, using natural units would induce changes in the dimensions of many physical quantities commonly used in our day-to-day life, such as speed, energy and power. We will indeed see that energy would have the same dimension as mass, and power as a mass per meter. These changes are far too important to be acceptable in real life. However, from a theoretical standpoint, the "natural" unit system is important and useful: it simplifies the physical relations and reveals interesting physical aspects.

3.5 Special Relativity in the Real World

Until now we have considered fixed objects in inertial frames, but in real life, we inevitably encounter accelerations and curves; hence, it is necessary to see how Special Relativity copes with these situations. We will first see the Langevin's famous twin paradox and discover surprising results. We will next address the general case of accelerated trajectories and see the additional postulates that are required. We will then demonstrate the Lorentzian distance triangle inequality and finally draw some fundamental consequences regarding the law of inertia and the notion of time.

3.5.1 Langevin's Famous Paradox

3.5.1.1 Langevin's Famous Twin Paradox

Two twins, named M (for Mobile) and F (for Fixed), are together in a city named A. Then the twin M leaves F and undergoes a trip into space. He first goes to a planet B, and as soon as he reaches B, he immediately returns to the city A (on Earth) and meets his twin F who stayed there all the time.

For the sake of simplicity, we will assume that M travels from A to B along a straight line and at the constant velocity V, and that he also returns from B to A along the same straight line and with the constant velocity W. We will also assume that the Earth is an inertial frame.

Question: Will they be the same age at the return of the traveler twin M? If not, who will be younger?

When the twin M goes from A to B, the twin F can apply the time dilatation law, which says that the proper time duration incurred by M, denoted by τ_{AB}^M, is seen dilated on Earth. Thus: $T_{AB}^F = \gamma_V . \tau_{AB}^M$, with T_{AB}^F being the time duration seen on Earth between the departure of M and his arrival at B.

As the twin F is fixed in the Earth frame, T_{AB}^F is identical to his proper time duration from A to B, denoted by τ_{AB}^F. We thus have:

$$\tau_{AB}^F = T_{AB}^F = \gamma_V \tau_{AB}^M. \tag{3.8}$$

This relation shows that τ_{AB}^F is always greater than τ_{AB}^M since γ_v is always greater than 1.

Then, when the twin M returns from B to A: the same[4] time dilatation law applies, implying that the proper time of the mobile twin M is again seen dilated by F, so that we have:

$$T_{BA}^F = \tau_{BA}^F = \gamma_W \tau_{BA}^M. \tag{3.9}$$

When they meet again, the total proper time duration incurred by F is:

$$\tau_{total}^F = \tau_{AB}^F + \tau_{BA}^F.$$

Considering (3.8) and (3.9), we deduce that: $\tau_{total}^F > \tau_{AB}^M + \tau_{BA}^M$, so that finally:

$$\tau_{total}^F > \tau_{total}^M \blacksquare \tag{3.10}$$

The proper time of an individual is the biological time incurred by his body: for instance, if your heart beats 60 times per minute, then your proper time is given by your number of heart beats. Hence, the relation 3.10 shows that the twin F is older than the twin M when they meet again.

However, this result faces a serious objection: the principle of Relativity implies that the same scenario can be considered from the perspective of the mobile twin M: he can consider himself as fixed, whereas it is his twin F who is moving together with the Earth frame. He can then reason the same way, which will lead to the opposite conclusion: F will be younger when they meet again, since F moved (together with the Earth frame), whereas M remained fixed.

Historically, this was a serious issue for the theory of Relativity in 1911, but a clear solution was found: Actually, the situations of F and M are not symmetrical because the frame of F is inertial (by hypothesis), whereas the one of M cannot be inertial since M necessarily incurs accelerations: first when he leaves Earth, then when he makes a U-turn at B, and finally when he lands on Earth.

The laws of Relativity were indeed demonstrated in the context of inertial objects in inertial frames. This scenario clearly shows that one cannot automatically extend these laws to non-inertial frames. However, we will see that the inertial twin M is allowed to apply the laws of Relativity, whereas the accelerated one F is not. It is then the inertial twin M who is right: he will be older than his traveling twin when they meet again.

This scenario is quite counterintuitive: the further the traveling twin goes, the younger he will be compared to his inertial twin. Theoretically, he can even return on Earth many centuries after leaving. Nevertheless, he won't be younger than when he left: his aging process always continues at the same pace. He notices nothing special regarding time in his spaceship: it still takes 4 minutes to make a boiled egg the same way he likes them at home. These aspects will be further explained in the next chapter.

3.5.2 The "Clock Postulate" or the Proper Time Universality

The clock postulate states that **the proper time of a perfect clock is independent of its acceleration**.

This postulate is based on experiments where some very high-precision clocks were submitted to very high accelerations, up to $10^{18}g$, and their proper times remained unchanged. These clocks were still beating at the same pace, as seen by a co-located observer. Of course, not all clocks are able to stand such accelerations without altering their periods (or even breaking), but the fact that some exist induced us to generalize these observations with this clock postulate.

This also means that **the proper time is universal**, preferring the word "universal" to "absolute" in order to avoid possible confusions with the classical Newtonian space and time which were assumed to be absolute. We saw that the principle of Relativity implies that the proper time is independent of the speed of the clock carrying this time, but without the clock postulate, we could not have said that the proper time is also independent of the acceleration.

This important proper time universality property will be confirmed in General Relativity in the context of gravitational fields (cf. Section 6.2).

3.5.3 The "Inertial Tangent Frame Law" (ITF Law)

We will first define the inertial tangent frame and then present the inertial tangent frame law:

3.5.3.1 The Inertial Tangent Frame (ITF): Definition and Main Properties

We recall that a curved (or accelerated) trajectory, as any trajectory, is a succession of events characterizing the space-time position of a mobile object. At each event of its trajectory, we can consider the inertial frame that has the same velocity as the object at this event: this defines the inertial tangent frame (ITF) (Figure 3.9).

FIGURE 3.9
Accelerated trajectory.

During a very small time lapse around this event, the object speed tends to zero in the ITF; it is indeed assumed that the speed of a real object is a continuous function. Consequently, the time t of the ITF is the same as the object proper time τ during this very small time lapse, since: $dt = \gamma_v \cdot d\tau$.

The object acceleration (if nonzero) is not null in the ITF, and it is the only privileged direction in the ITF.

3.5.3.2 The "Inertial Tangent Frame Law" (ITF Law)

The inertial tangent frame law (ITF law) states that an inertial observer can consider an accelerated (or curved) trajectory as a succession of infinitesimal sections, and apply in each section during a small time lapse the laws of Relativity that were established between inertial frames, between his frame and each inertial tangent frame. He can then integrate the results obtained in this succession of infinitesimal sections along the trajectory of the object.

One should pay attention that the frame K of the observer who considers the accelerated object must be an inertial frame. We indeed saw with Langevin's famous twin paradox that the twin who incurred accelerations was not allowed to apply the time dilatation law.

Remark 3.7: The ITF law relies on the clock postulate and the assumption that the space-time universe and objects trajectories are differentiable.

The following example illustrates this ITF law:

3.5.3.3 Langevin's Travelers Paradox with a Differentiable Accelerated Trajectory

In Figure 3.10 below, the mobile twin M has an accelerated (curved) trajectory, whereas his twin F has an inertial one at the constant velocity V relative to an inertial frame K.

FIGURE 3.10
Accelerated and inertial trajectories.

The above ITF law says that F can consider that M incurs a succession of infinitesimal inertial sections, and that he can apply during a small time lapse the laws of Relativity between his inertial frame and each inertial tangent frame along M's trajectory. The twin F is thus allowed to apply the time dilatation law during each small time lapse: $d\tau_F = \gamma \, d\tau_M$ (the time of F's frame being also F's proper time since he is fixed in his frame).

Then, F can integrate these results, meaning that the time duration between A and D incurred by him is: $\tau_F^{AD} = \int_A^D \gamma_v \, d\tau_M$. Consequently, we have at the arrival at the point D: $\tau_F^{AD} > \tau_M^{AD}$, meaning that the inertial twin will be older than his accelerated twin.

This result is actually the same as the previous twin paradox: this Travelers paradox scenario can indeed be considered from an inertial frame K' moving at the speed V relative to K. The twin F is then fixed in K', and M makes a loop; hence, this scenario in K' is identical to the twin paradox.

It would be wrong to deduce that the proper time of the accelerated twin M has flowed more slowly than the one of the inertial one F because M is younger than F when they meet at D. The proper time of the twin M kept flowing at the same pace, in accordance with the clock postulate. Their time difference at D is due to de-synchronization effects[5].

3.5.4 Implication on the Fundamental Law of Inertia

We saw that for the inertial traveler: $c.\tau_F^{AD} = \sqrt{\Delta s^2(AD)}$. Regarding the accelerated traveler, M, the duration of his trip from A to D, measured with its proper time, is the curvilinear integral along his accelerated trajectory: $c.\tau_M^{AD} = \int_A^D \sqrt{ds^2(dM)}$. This summation is **a route-dependent quantity**[6], with its maximum being the inertial trajectory between A and D.

This shows that the classical law of inertia must be revised since the spontaneous trajectory of an object submitted to no force (inertial) still is a straight line, but the latter maximizes the (Lorentzian) distance between two points. Hence, the law of inertia must evolve and becomes:

An inertial object going from a point A to a point B chooses the trajectory that maximizes[7] the (Lorentzian) distance covered, which is also the time spent and measured with its proper time.

This formulation will be especially important in General Relativity as it will enable us to find the trajectories of inertial objects in a non-Euclidean space, where there are no straight lines.

3.5.5 Langevin's Travelers Paradox and the Relativistic Triangle Inequality

Let's now consider the following scenario where the twin F undergoes an inertial travel from A to C, whereas his twin M makes a first inertial travel from A to B, and then an instantaneous change of direction toward C followed by an inertial travel from B to C. Then finally, F and M simultaneously arrive at C. *Will both twins be the same age when they meet at C* (Figure 3.11)?

FIGURE 3.11
Sequence of inertial trajectories.

The twin F, being inertial, is allowed, thanks to ITF law to apply the time dilatation law and to make the summation from A to C: he thus sees the proper time taken by his twin M dilated in both sections from A to B and from B to C, meaning: $\tau^F_{AB} = \gamma_V \tau^M_{AB}$ and $\tau^F_{BC} = \gamma_W \tau^M_{BC}$. Consequently: $\tau^F_{AC} > \tau^M_{AB} + \tau^M_{BC}$, meaning that M is younger than F when they meet again at the point C.

This scenario provides us with a simple way to demonstrate the Lorentzian norm triangle inequality:

3.5.5.1 The Triangle Inequality of the Lorentzian Distance

Consider now three events: A*, B* and C*: we saw that $\sqrt[2]{\Delta s^2(A*B*)}$ is the proper time duration of an inertial travel from A* to B* (multiplied by c^2); and the same for $\sqrt[2]{\Delta s^2(B*C*)}$ and $\sqrt[2]{\Delta s^2(A*C*)}$.

The above Langevin's Travelers scenario shows that the proper time duration for the inertial traveler from A* to C* is greater than the sum of the one from A* to B*, and the one from B* to C*, which translates to:

$$\sqrt[2]{ds^2(A*C*)} > \sqrt[2]{ds^2(A*B*)} + \sqrt[2]{ds^2(B*C*)} \ \blacksquare \qquad (3.11)$$

This is the opposite from the classical triangle relation whereby in any triangle we have: AC < AB + BC.

Remark 3.8: This triangle inequality relation requires that the space-time intervals are time-like and that B* is in the future of A*, and C* in the future of B*.

Remark 3.9: Given the importance of this triangle inequality, two other demonstrations are presented in Volume II.

3.5.6 Fundamental Implications on the Notion of Time

3.5.6.1 Implication on the Notion of Time: How to Synchronize Time?

The previous scenarios show that if we want to synchronize the time in the whole inertial frame K, we cannot do it by the simple and intuitive method which consists of building a factory of perfect clocks in a point A, giving all the same official time. Then, we bring to any point B of the frame K one clock from this factory, and it would give the time of the frame K in B.

The previous scenario shows that this method does not work because the clock will inevitably incur accelerations during its travel from A to B, which will make its time advancing slow compared to the time that would give an inertial clock moving from A to B[8]. Besides, an inertial clock moving from A to B displays when arriving in B a time which is behind the time of the inertial frame K[9].

Consequently, the time synchronization in an inertial frame can only rely on the method presented in Section 1.3.2.2, i.e., by broadcasting the time information from one point, and then adding the corresponding signal transmission delay. However, this method requires to have absolutely straight lines, which is not a problem in Special Relativity, but it is in General Relativity as we will further see, and hence in the real world. The consequence will be the impossibility to have a synchronized time throughout the whole universe.

3.6 Questions and Problems

Problem 3.1: Relativity of Present, Past and Future, and Impossibility of Speeds Greater than c

Consider two events E* and F* that are simultaneous in an inertial frame K. For the sake of simplicity, we set the origin-event of K to be E*. We place the OX axis of K such that F* is on this axis, with its abscissa being 1.

Q1A: Show that there exist several inertial frames where the event F* is seen occurring after the event E*, and several other frames where F* is seen occurring before the event E*.

Q1B: What distinguishes the two categories of frames?

Q2: Let's have a frame K′ which goes at the speed v along the direction FE in K, and with v < c. In K′, an object M has been seen going from E* to F* at the constant speed s. Find the value of s, and compare it with c.

Q3: In another frame K″ that goes at the speed v along the direction EF in K, is the previous mobile object M seen arriving at the event F* after or before having left the event E*? Please comment.

Problem 3.2: The Special Inertial Frame associated with any Pair of Events

Consider any pair of events (E, F) with their coordinates being in an inertial frame K: E(Xe, Te), and F(Xf, Tf), with Tf > Te and Xf > Xe. Use Minkowski diagrams to show that:

Q1A: If EF is a time-like interval, there exists an inertial frame K′ where they are co-located. Show this frame K′ in the diagram. Hint: choose E as the common origin-event of K and K′.

Q1B: What is the angle made by the frame K′ with K?

Q2A: If EF is a space-like interval, there exists an inertial frame K″ where they are simultaneous. Show this frame in diagram.

Q2B: What is the angle made by the frame K″ with K?

Problem 3.3: Twin Paradox with the Simultaneity Lines, Part 1

Alice is staying on Earth (frame K, assumed to be inertial), whereas Bob goes to the cosmos at the constant velocity $v = 0.866c$; the corresponding gamma factor is: 2.00. Both Alice and Bob set their watches at zero when Bob leaves. (We assume that Bob already goes at the speed $v = 0.866c$ at the time $t = t′ = 0$.) Let's call F* the event: (location: Bob; time: 10 years on Bob's watch).

Q1: Make a Minkowski diagram with K being the Earth frame, and Alice at the point O. The distances are expressed in light-years; the times in years. Calculate the angle α of the Minkowski diagram; mark the event F* and give its coordinates in both K′ and K.

Q2A: Bob, after 10 years of his trip, wonders how much time it represents for Alice. He thus imagines an event G* which is: "location: Alice; time: simultaneous with F*" (the simultaneity being considered by him). Represent the event G* in the Minkowski diagram.

Q2B: Calculate the time Tg of G* in Alice's frame (K).

Q2C: What is the ratio Tf/Tg? Express this ratio as a function of the gamma factor, and give its value.

Q3: Is this consistent with the result saying who is the older when Bob returns? If there is a discrepancy, explain the reason, and how to solve it.

Problem 3.4: Twin Paradox with the Simultaneity Lines, Part 2

This problem is a continuation of the previous one. At the event F*, Bob instantaneously makes a U-turn, and then returns to see Alice on Earth. His trip back is at the same constant speed of v = 0.866c. The event "Bob meets Alice at his return" is denoted by M*.

Q1A: In the previous Minkowski diagram, draw Bob's return trajectory and mark the event M*.

Q1B: Immediately after having made his U-turn, Bob sends a light pulse toward Alice. Draw the trajectory of these photons, which we will call the "new line of light".

Q1C: Draw the new line which corresponds to all events that Bob considers simultaneous with him immediately after his U-turn. This line will be denoted by F* − X".

Q2A: Immediately after having made his U-Turn, Bob asks himself the same question as the previous Question 2A: What time is it for Alice now, the simultaneity being relative to Bob? He thus imagines an event N* defined as: "location: Alice; time: simultaneous with Bob immediately after F*". Mark this event N* in the diagram.

Q2B: Give the coordinates of N* in K. Please comment.

Q3A: What is the travel time for Bob's return trip, seen by Bob? What is the total travel time, seen by Bob? What will Bob think during his way back about his age compared with Alice's when he returns?

Q3B: What is the travel time seen by Alice? Who is the older at Bob's return?

Q3C: How can we explain the age difference: Can we consider that Bob's proper time has flowed more slowly?

Q4: Let's now imagine the same scenario, but with Bob staying at the point O of K, and Alice being at O' of K'. Will the Minkowski diagram be the same as previously if we swap Bob's and Alice's names? What result will it give concerning the age difference between Bob and Alice? Please comment.

Problem 3.5: Demonstration of the Lorentz transformation with Natural Units

We will demonstrate the Lorentz transformation with natural units, in the same way as we did in Section 2.2.3, i.e., using two scenarios giving four equations enabling us to obtain the four parameters of the Lorentz transformation. The inertial frames K and K′ share a common origin-event O, and K′ is going at the speed v, expressed in meter/second, along the OX and OX′ axes. We place a fixed clock at the origin point O of the frame K. At the time t=0 in K, this clock emits a flash, which constitutes the origin-event O* of both frames. Then, at the time t=1 second in K, the clock emits a second flash, which constitutes the event E*.

Q1A: Express the coordinates of E* in K with natural units. We will express time and distance with the distance unit, which is the meter. The coordinates in natural units of E* in K will be denoted by T_e and X_e.

Q1B: Express the coordinates of E* in K′ with natural units. These will be denoted by T_e' and X_e'.

Q1C: The Lorentz transformation has four parameters denoted by a, b, m and n, which are such that: $X_e'=a.X_e+b.T_e$ and $T_e'=m.X_e+n.T_e$. Use the results of the previous questions to find the parameters b and n.

Q2A: We place a fixed clock at the origin point O′ of the frame K′. At the time t′=1 second in K′, this clock emits a flash, which constitutes the event F*. Express the coordinates of F* in K′ in natural units.

Q2B: What are the coordinates of F* in K in natural units?

Q2C: Use the results of the previous questions to find the parameters b and n. Then, write the Lorentz transformation in natural units with the matrix format.

Problem 3.6: Synchronization Impossibility by Moving a Clock

In an inertial frame K, we want to synchronize the times in two points A and B, both fixed in K. To synchronize the time of B with the one of A, we can think of bringing a clock from A to B, this clock being identical to the one of A. We will show that this method does not work, and that the clock arriving in B will display a time which is different from the one of a clock in B that is synchronized with the one of A.

We build a clock C which is absolutely identical to the one in A. We move this clock C, traveling at constant velocity along the straight line AB. Let's call K′ the inertial frame of the clock C. The origin-event O′ of K′ is: "C is at the point A". The O′X′ axis of K′ is along the segment AB.

Q1A: When the clock C arrives at B, does it display the same time as a clock in B which is synchronized with the time of the clock in A?

Q1B: If it is not synchronized, will the clock C be ahead or behind the clock B?

Q2A: We build another clock D which is absolutely identical to the one in A. This clock D now has an accelerated trajectory from A to B. Will the clock D, when arriving at B, display a time which is ahead or behind the clock C when it is at B (as in the previous question)? You may use the qualitative result of Langevin's Travelers paradox.

Q2B: Will the clock D, when arriving at B, display a time which is ahead or behind the time of the frame K?

Problem 3.7: The Clock in a Satellite

A satellite is at an altitude of 100,000 km from the Earth surface and orbits the Earth in 1 hour. We have placed a perfect clock inside this satellite, which is identical to perfect clocks on Earth giving the official second. Does an observer on Earth see the satellite clock beating more slowly or more rapidly than an identical perfect clock on Earth? By what percentage? We assume that the Earth is inertial.

Problem 3.8: Two Opposite Light Beams

You have two lasers that you orientate in opposite directions. You trigger their beams simultaneously. The front ends of these beams are, respectively, denoted by A and B. At what speed is the length segment AB increasing? Is this contradictory with the existence of a maximum speed?

Problem 3.9: Light Beam and a Moving Mirror

In a frame K, a light beam hits a mirror which is moving along the OX axis. This beam makes an angle α with the mirror.

Q1: What angle does the beam make after its reflection by the mirror? Hint: consider this scenario from the frame K' where the mirror is fixed.

Q2: In K', what angle does the beam make with the mirror?

Problem 3.10: Velocity Composition and the Triangle Inequality.

In an inertial frame K', we have three fixed points A, B and C such that AB=BC; BC is perpendicular to AB; and AB is parallel to O'X'.

An object is going at the speed s=c/2 from A to B, then turns instantaneously toward C and goes from B to C at the same speed s=c/2. The object velocities from A to B, and from B to C are, respectively, $\vec{W_1}$ and $\vec{W_2}$ (Figure 3.12).

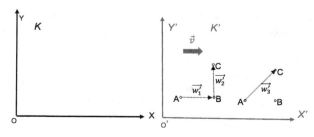

FIGURE 3.12
Different velocity compositions from A to C.

The frame K′ is going at the speed v relative to an inertial frame K. We have: v=0.75 c, which corresponds to the gamma factor of: $\gamma = 1.51$.

Q1: Calculate the velocity $\overrightarrow{W_1}$ of the same object seen from the frame K, between A and B.

Q2: Calculate the velocity $\overrightarrow{W_2}$ of the same object seen from the frame K, between B and C.

Q3: In the frame K′, we have a second object which goes directly from A to C at the velocity $\overrightarrow{W_3'}$. This second object leaves the point A at the same time as the first object, and both arrive simultaneously at C.

What is the speed W_3'? What is its component parallel to OX, denoted by W_{3x}'? What is its component parallel to OY, denoted by W_{3y}'?

Q4: Calculate the velocity $\overrightarrow{W_3}$ of the second object seen in the frame K, between A and C. Calculate first its component parallel to OX, denoted by $\overrightarrow{W_{3x}}$.

Q5: Calculate the component parallel to OY of the velocity $\overrightarrow{W_3}$, denoted by $\overrightarrow{W_{3y}}$.

Q6A: The distance AB is 1 light-year in K′, and we still have AB=BC. What is the time in K′ taken by the two objects to go from A to C?

Q6B: What is the time in K taken by the two objects to go from A to C?

Q7A: What is the time taken by the second object to go from A to C, measured with its proper time?

Q7B: How does it compare with the one taken by the first object? Give a first qualitative answer.

Q7C: Make an exact calculation; we give: $\gamma_{0.35c} \approx 1.067$ and $\gamma_{0.5c} \approx 1.155$.

Notes

1 We recall that the light speed invariance postulate concerns inertial frames. In some accelerated frames, it is possible to see light going faster than c.

2 The reasons why the Minkowski diagram functions like this are explained in Volume II Section 2.3.1.

3 Considering two norms which are proportional to be the same, as shown in Volume II Section 2.1.1. In particular, we could have defined $ds^2 = (x^2 + y^2 + z^2) - c^2t^2$, and indeed, it is defined so in some domains of physics, or in some countries.

4 Cf. Section 2.1.2.5.

5 Cf. more with Problems 3.3, 3.4 and Volume II Section 2.7.2.1.

6 These words are from H. Bondi.
7 The word "extremizes" can be used instead of "maximizes" because the Lorentzian distance can be defined as the opposite of the one used in this document.
8 Cf. more in Volume II Section 2.6.2.
9 Cf. Problem 3.6.

4

The Next Revolution: Dynamics

Introduction

Having seen that the classical kinematics laws are wrong, it is logical to check whether the classical physical laws remain valid. Hence, this chapter begins with an overview of the main classical laws of dynamics. Then, the compliancy of the momentum with the principle of Relativity will be checked, and the outcome will be negative. The consequences are huge since all classical dynamic laws rely on the momentum. We will then search for the correct relativistic Momentum: the core of its principle will be examined and adapted to the principles of Relativity, using in particular the concepts of Minkowski. The important contravariance property of the relativistic Momentum will then be demonstrated, and subsequently used to check that the new Momentum definition respects the principle of Relativity. We will then examine the important consequences of the new Momentum.

4.1 What's Wrong with the Main Classical Laws?

4.1.1 Overview of the Main Classical Laws

4.1.1.1 The Newtonian Force

The famous Newton's second law states that an object of *inertial* mass which incurs an acceleration $\vec{a}$ relative to an inertial frame is subjected to the force which is: $\vec{f} = m.\vec{a}$.

We can first see that if no force is applied on an object, it continues with a constant velocity and along a straight line, which is the law of inertia (also called the first Newtonian law).

We can also see that for a given change of speed (acceleration) of an object, the greater its inertial mass, the greater the force required, hence the name *inertial* mass.

DOI: 10.1201/9781003201335-4

Remark 4.1: This second Newtonian law actually is a postulate, as is the law of inertia.

An important property of the Newtonian force is its vectorial additivity: if an object is submitted to several forces, the combination of these forces results in their vectorial sum: $\vec{F} = \sum_{i=1}^{n} \vec{f_i}$.

4.1.1.2 Case of Non-Inertial Frames: Inertial Forces

If we are in a non-inertial frame K′, there is an inertial force which tends to keep us inertial, hence the name "inertial force". For instance, if we are inside a lift incurring an acceleration upward, we can feel a force pushing you downward. More generally, the inertial force is: $\vec{F_i} = -m.\vec{a}'$, where $\vec{a}'$ is the acceleration of the frame K′ relative to any inertial frame.

A particular case is when K′ is the frame of the accelerated object. In its own frame, the object speed and acceleration are always null. This translates to the equality between the inertial force incurred by the object and the Newtonian force to which it is submitted, and which is $\vec{F_n} = m.\vec{a}'$. We indeed have: $\vec{F_n} + \vec{F_i} = 0$.

4.1.1.3 The Gravitational Force

The gravitational force is: $F = m_1 m_2 \dfrac{G}{R^2}$. Many experiments showed that when you drop any object, it falls at the same speed and acceleration whatever its gravitational mass and its inertial mass, provided that the friction with the air is neglected. Consequently, we have: $\dfrac{m}{m_g} = \dfrac{g}{a} = \text{cst}$ whatever the mass. For the sake of simplicity, this constant was set to be 1; hence: $m = m_g$ and $g = a$.

A principle of classical physics states that the mass of any object remains constant: an object can incur transformations of many kinds (by thermal, mechanical, chemical means, etc.), but the total mass of the resulting pieces is equal to the mass of the initial object.

4.1.1.4 The Momentum

The momentum of an object was defined as: $\vec{p} = m.\vec{v}$, with the speed being considered from an inertial frame. There is a direct link between the momentum and the Newtonian force since we have: $\vec{f} = \dfrac{d\vec{p}}{dt}$.

We can see that the momentum is a vector which is always tangent to the object trajectory, and proportional to both its speed and its mass. If there is no external force, the momentum derivative is null; hence, it is constant, meaning that the velocity vector is constant. The inertia law can thus be stated as follows: the momentum of a body subject to no force is constant (considered from an inertial frame).

In classical physics, the momentum concept is especially meaningful in the case of systems composed of several parts. Indeed, the momentum of such system, defined as the vectorial sum of the momenta of all its parts, remains constant if the system is not submitted to any external force. (Such a system is said to be isolated or closed.) We thus have: $\vec{P}_{system} = \sum \vec{p}_i = \sum m_i \vec{v}_i = \overset{®}{cst}$.

The conservation of the system momentum is a remarkable law that proves to be very useful in various contexts such as collisions between objects, the push of rocket engines, the backward movement of cannons, etc.

4.1.1.4.1 Reciprocity of Forces

This fundamental system momentum conservation law also explains the principle of equality between the action and the reaction[1], which is the third Newtonian law: when a body exerts a force on a second body, the second body simultaneously exerts a force on the first body, and these forces are equal in magnitude and opposite in direction. An example is the force of gravity. The demonstration is the following:

> For the sake of simplicity, let's consider from an inertial frame K an isolated system having two objects only, A and B, each one exerting a force toward the other. The momentum conservation law gives: $\vec{P}_a + \vec{P}_b = \overline{cst}$.
>
> The constancy of this expression implies that its derivative relative to the time t is null: we thus have: $\dfrac{d\vec{P}_a}{dt} + \dfrac{d\vec{P}_b}{dt} = \vec{0}$; hence: $\vec{F}_a = -\vec{F}_b$.

From an axiomatic perspective, the Momentum conservation appears to be the root from which the force is derived, as it explains the force reciprocity and the reason why the force only affects the second derivative of the position of an object in an inertial frame.

4.1.1.5 The Concept of Energy

Leibnitz showed that the deformations made by a body in motion when crashing are proportional to mv^2 (the body's mass and to the square of its speed), a quantity he called "living force".

Then the Dutch scientist C. Huygens showed that in case of elastic[2] collisions between objects (for example, billiard balls), their speeds after the collision can be calculated using two laws: the system momentum conservation law (cf. above) and a new conservation law concerning the sum of the mv^2 of all the parts. He called this quantity mv^2: "the energy of motion", which coincided with Leibnitz's "living force".

Finally, the kinetic energy was defined as $\frac{1}{2} mv^2$, which still was consistent with Leibnitz's and Huygens's findings, mainly because it matched with the two following important findings:

The first law of thermodynamics, found by R. Clausius in 1850, states that the total energy of an isolated system remains constant, whatever the transformations of any kind occurring within such a closed system, and involving exchanges between different sorts of energy: kinetic, thermal, electromagnetic, chemical, etc. In particular, it has been observed that in the cases of exchange between thermal and kinetic energies, it is the variation of ½ mv² which matches with the variation of the thermal energy of the system.

The French scientist G. Coriolis found the relation between the force and the energy given by this force. He defined the "work of the force" as follows, and showed that it equals the variation of the kinetic energy of the object submitted to this force. When a force $\vec{F}$ makes an object move from A to B, the work of this force is defined as: $W = \vec{F} \cdot \overrightarrow{AB}$. We have assumed the force is constant and parallel to the object motion; if it is not constant, we have:

$$W_{AB} = \int_A^B F\,dx = \int_A^B m\frac{dv}{dt}dx = \int_A^B m\frac{dx}{dt}dv = \int_A^B mv\,dv$$

$$= \int_A^B \frac{1}{2}m\,dv^2 = \left[\frac{1}{2}mv^2\right]_A^B = \frac{1}{2}mv_B^2 - \frac{1}{2}mv_A^2 \qquad \blacksquare$$

We thus have: $W = \Delta E = \Delta(1/2mv^2)$, meaning that the work of the force is equal to the kinetic energy variation. Note that the energy required to go from a state A to a state B can be achieved either by a small force covering a long distance, or by a stronger force on a shorter distance. This law has many applications in mechanics, for instance the leverage effect and the gear box.

> In the even more general case where the force is not parallel to the movement, but makes an angle α with it, the above expression $W = \vec{F} \cdot \overrightarrow{AB}$ is equal to: $W = F.AB.\cos\alpha$.

Another interest of the work of the force is its link with the concept of potential energy which proved to be useful in various domains: gravitation, electromagnetism and even nutrition.[3]

***

It is remarkable how simple were these fundamental laws, and this simplicity together with the accuracy of their results induced scientists to believe they were absolutely true. There is indeed a long tradition that the simplicity of a theory and even its beauty speak in favor of its validity. However, Einstein showed that all these laws were wrong, and he replaced them with new laws forming an even simpler theory, from an axiomatic perspective.

4.1.2 The Classical Momentum Definition Is Not Acceptable in Relativity

We will first address the momentum as it appears to be the root from which the force and the energy are derived. (We indeed saw that the force is the derivative of the Momentum, and the energy is the work of the force.)

The importance of the momentum stems from its fundamental system conservation law: if a system is isolated, its momentum is conserved (constant). We will show with a simple scenario that this fundamental physical law does not respect the frame equivalence postulate, which is not acceptable.

In an inertial frame K, two identical cars A and B move at the same speed V but in opposite directions so that they incur a collision. After this crash, both cars remain stuck together without moving. This collision caused deformations and some heat dissipation; hence, it is said to be inelastic, and even completely inelastic because both cars are fixed in K after the collision, meaning that all the initial kinetic energy has been transformed into other sorts of energy (Figure 4.1).

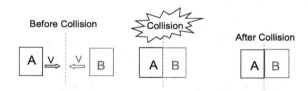

FIGURE 4.1
An inelastic collision.

The system momentum conservation law in the frame K means that the system momentum before the collision, denoted by SM1, must be equal to the system momentum after the collision, denoted by SM2.

We have: $SM1 = mV + m(-V) = 0$, with m being the mass of each car.

After collision, we have: $SM2 = m0 + m0 = 0$. We can see that the system momentum conservation law is respected in the frame K.

Let's now consider this same scenario from another frame K' moving at the speed w relative to K along the X axis (the cars' line motion). We want to check if the system momentum conservation law is respected in K', and for the sake of simplicity, we choose $w = V$.

Let's first calculate in K' the system momentum before the collision, denoted by SM1': the speeds of A and B in K' before the collision are denoted by V1a' and V1b'.

We thus have: $SM1' = m.V1a' + m.V1b'$.

We have: $V1a' = 0$ since K' is moving at the speed V relative to K.

Regarding V1b': it is the result of the composition of the speed $-V$ of B relative to K, and the speed $-V$ of K relative to K': $V1b' = (-v) \oplus (-v)$, which gives: $V1b' = -2V/(1 + V^2/c^2)$. Hence: $SM1' = -2mV/(1 + V^2/c^2)$.

Let's now calculate the system momentum after the collision in K', denoted by SM2': both cars are moving at the speed –V relative to K' since they are not moving within K; we then have: SM2' = mV2a' + mV2b' = **–2mV**.

We can see that SM2' is always different from SM1', which means that the system momentum conservation law does not respect the principle of Relativity, even if the difference is minor when v << c. This is not acceptable since it must be possible to apply any physical law from any inertial frame: no frame is privileged. Consequently, either the system momentum conservation law is wrong, or it is correct, but we used an incorrect momentum definition. The latter possibility is favored because the system momentum conservation appears to be a very fundamental law.

We need then to find the correct momentum definition.

4.2 The Momentum in Relativity

The relativistic Momentum is the cornerstone of dynamics; hence, we will examine this notion carefully. A physical analysis is presented afterward, and several demonstrations will be given, revealing different aspects and with different levels of complexity[4].

Preliminary remark: The relativistic Momentum will be denoted with the capital P and the classical momentum with p.

4.2.1 Searching for the Relativistic Momentum Definition

Extending Einstein's finding that time and distance were relative and linked together, Minkowski showed that we are in a 4D space-time universe, with the elementary concepts being the event and the proper time. Consequently, physical laws relating to objects' motions are 4D by essence, and it is from these 4D laws that the classical 3D ones are derived. The latter are extremely good approximations for speeds low compared to c, as in the case of momentum. We must then seek the 4D law from which the classical 3D momentum, p = mv, is derived. We just saw that the velocity v used in the momentum definition was not appropriate; hence, we will investigate the notion of velocity in Relativity.

4.2.1.1 The General Velocity Definition in Relativity

We saw that in Relativity, the space-time evolution of a point object M is characterized by a succession of events. Hence, the general notion of velocity is the time rate of change of the events characterizing this object along its trajectory.

However, Relativity showed that time is not an absolute: in particular, the time of an object which is moving with respect to an inertial frame K is seen dilated in K: it constitutes a natural source of time which is seen in K beating more slowly than the time of K. Hence, there are as many natural sources of time as there are moving objects. In addition, the frame K has its time, but actually this time is an artificial construction of men, based on the proper time in one fixed point of K, and then expanded to the whole frame using exchanges of signals to ensure time synchronization (as we saw in §1.3.2).

Let's then see how the 4D velocity is defined once the source of time is given. The time of this source will be denoted by θ. The 4D velocity, $\vec{S}$, of a point object M is the logical 4D extension of the classical 3D velocity: at each time θ, the point object M is characterized in K by the event: $M^K(\theta) = [T(\theta), X(\theta), Y(\theta), Z(\theta)]$. Then, at the time $\theta + d\theta$, this object has moved from the event $M(\theta)$ to the event $M(\theta+d\theta)$. The small space-time separation vector $[M(\theta), M(\theta+d\theta)]$ is denoted by $\overrightarrow{dM(\theta)}$, and the 4D velocity associated with the time source θ is:

$$\overrightarrow{S(\theta)} = \frac{\overrightarrow{dM(\theta)}}{d\theta} = \left[\frac{dT(\theta)}{d\theta}, \frac{dX(\theta)}{d\theta}, \frac{dY(\theta)}{d\theta}, \frac{dZ(\theta)}{d\theta} \right].$$

We then have to find the appropriate source of the time involved in the 4D velocity used in the definition of the Momentum of a given point object M.

4.2.1.2 Seeking the Appropriate Time Source for the Relativistic Momentum

To find the appropriate time source for the velocity, we will base our reasoning on the physical properties that the relativistic Momentum must have. Let's then recapitulate these:

A. The relativistic Momentum must be a 4D quantity that has the form: $\overrightarrow{P(\theta)} = m. \overrightarrow{S(\theta)}$. The scenario of §4.1.2 indeed showed that there is an issue with the classical speed, but not with the mass, and we will assume that the relativistic Momentum remains proportional to the object mass m.

B. The object mass is intrinsic to the object and hence invariant, meaning identical in all inertial frames[5]. We have no reason a priori to invalidate this property; however, we will further see that some properties of the mass will evolve, in particular its constancy.

C. The Momentum of an isolated object remains constant, as seen from an inertial frame. This expresses the fundamental law of inertia.

D. The system Momentum, defined as the vectorial sum of the Momenta of all its parts, remains constant for an isolated system, again as seen from an inertial frame.

These last two properties are very fundamental physical laws which confer the great importance of Momentum. We saw in particular that the concept of inertial frame and the definition of its time rely on the law of inertia.

(The exchange of signals needs to be at constant velocity along straight lines.) Hence, it would not be logical to use the time of an inertial frame in the definition of the Momentum, but a more primitive physical time, meaning the proper time of a physical entity. Then, as this physical entity needs to be related to the object M, it is logical to deduce that the source of the time to be used for the Momentum of the object M is its proper time, denoted by τ. Thus, the Momentum of a point object should be:

$$\overrightarrow{P(\tau)} = m.\frac{\overrightarrow{dM(\tau)}}{d\tau} \blacksquare \tag{4.1}$$

This reasoning is by induction, which is common in physics; hence, we need to check the validity of its result, which we will do now. In addition, a mathematical demonstration is given in Volume II § 3.6.4, showing that there is no other possibility for the time source of the Momentum than the object' proper time.

The velocity $\vec{S}$ used in the Momentum, $\vec{S}(\tau) = \dfrac{\overrightarrow{dM(\tau)}}{d\tau}$, is called the proper velocity (or Four-Velocity or even Four-Speed) since it uses the object proper time. We thus have:

$$\overrightarrow{P(\tau)} = m.\vec{S}(\tau). \tag{4.2}$$

4.2.2 Momentum Validation and its Contravariance Property

4.2.2.1 Validity Conditions

Any physical law must be confirmed by experiments and observations. Hence, the first validity condition is that the spatial part of the relativistic Momentum is very close to the classical momentum for speeds which are far from c, since the classical momentum proved its validity in this context.

4.2.2.1.1 First Validity Condition

In an inertial frame K with its time denoted by t, an object is seen moving with the velocity $\vec{v}$. We must check that the classical momentum, $m\vec{v}$, gives similar results than the space part of the relativistic momentum for low speeds compared to c.

Let's then compare $\vec{v}$ and the space part of the proper velocity $\vec{S}$: The object' proper time is seen in K dilated by the γ_v factor: $dt = \gamma_v d\tau$. We can then express $\vec{S}$ in K:

$$\overrightarrow{S(\tau)} = \frac{\overrightarrow{dm(\tau)}}{d\tau} = \left[\frac{dt}{d\tau}, \frac{dx}{d\tau}, \frac{dy}{d\tau}, \frac{dz}{d\tau}\right] = \left[\frac{dt}{dt}\frac{dt}{d\tau}, \frac{dx}{dt}\frac{dt}{d\tau}, \frac{dy}{dt}\frac{dt}{d\tau}, \frac{dz}{dt}\frac{dt}{d\tau}\right]$$

$$= \gamma_v\left(1, \frac{dx}{dt}, \frac{dy}{dt}, \frac{dz}{dt}\right) = \gamma_v(1, \vec{v}).$$

Then, the relation (4-2) yields the relativistic Momentum expression in K:

$$\overrightarrow{P(t)} = m\frac{\overrightarrow{dM}}{d\tau} = m\vec{S} = m\gamma_v \left(\frac{dt}{d\tau}, \frac{dx}{d\tau}, \frac{dy}{d\tau}, \frac{dz}{d\tau} \right) = m\gamma_v \left(1, \vec{v} \right) \blacksquare \qquad (4.3)$$

We can see that when the magnitude of $\vec{v}$ is low relative to c, the spatial part of the relativistic Momentum, $m\,\gamma_v\vec{v}$, is very close to the classical momentum, $m\,\vec{v}$.

We will now address the other validation conditions, from both experimental and theoretical perspectives.

4.2.2.1.2 Other Validity Conditions

For high speeds even very close to c, many direct experiments were made within particle accelerators (although long after the discovery of Relativity), and all confirmed the new Momentum definition and its direct consequences.

From the theoretical perspective, the new law must be consistent with all postulates. Hence, we will first check its compliancy with the inertial frames equivalence postulate, since that failed with the classical momentum. We will then check its consistency with the Energy laws, and we will see that it reveals major novelties. We will finally check its consistency with the force.

To check the Momentum compliancy with the inertial frames equivalence postulate, we will first present an important property of the Momentum which is very useful in our further demonstrations: the Momentum is a contravariant 4-vector.

4.2.2.2 The Momentum Is a Contravariant 4-Vector

*We will show that when changing frame, the Momentum changes using the Lorentz transformation. Mathematically, this property is called **contravariance**.*

We will first show that the proper velocity, $\vec{S}(\tau) = \dfrac{\overrightarrow{dM(\tau)}}{d\tau}$, is contravariant, then the relation 4.2 implies that the Momentum is contravariant:

Let's consider a point object seen from two inertial frames, K and K′, with K′ moving at the speed w relative to K. The point object is characterized at the time τ of its proper time by the event $M(\tau)$. In K, this event occurs at the time t, and in K′ at the time t′. Then, at the time $(\tau + d\tau)$, the event $M(\tau + d\tau)$ is seen occurring at the time t + dt in K, and t′ + dt′ in K′.

The space-time separation vector $[M(\tau), M(\tau + d\tau)]$, denoted by $dM(\tau)$, is denoted in K by $\overrightarrow{dM^K(t)}$ and in K′ by $\overrightarrow{dM^{K'}(t')}$. Any space-time separation vector is transformed from K to K′ using the Lorentz transformation, denoted by (Λ_w). We then have: $\overrightarrow{dM^{K'}(t')} = (\Lambda_w)\,\overrightarrow{dM^K(t)}$.

In K′, the proper velocity is: $\overrightarrow{S^{k'}(t')} = \dfrac{\overrightarrow{dM^{K'}(t')}}{d\tau} = \dfrac{(\Lambda_w)\,\overrightarrow{dM^K(t)}}{d\tau} = (\Lambda_w)\,\overrightarrow{S(t)}.$

This relation shows that the proper velocity is contravariant since it changes from K to K' like any space-time separation vector. Hence, the proper velocity is also called "Four-Velocity" or "Four-Speed", the name "Four-X" meaning that the concerned vector X is contravariant.

We thus have:

$$\overrightarrow{P'(t')} = (\Lambda_w)\overrightarrow{P(t)} \blacksquare \qquad (4.4)$$

meaning that the Momentum is contravariant.

Remark 4.2: The name "contravariant" stems from the fact when changing frame, a fixed vector (such as a space-time separation vector) changes using the inverse matrix of the one used by the unit basis vectors. (Cf . more in Volume II § 4.1.1)

4.2.2.3 The System Momentum Conservation Law
Respects the Principle of Relativity

We will show that the fundamental system Momentum conservation law can be applied in any inertial frame while respecting the relativistic frame changing law regarding the velocity. Let's then examine the general case of collisions between two objects A and B, seen from two inertial frames K and K'. The figure below shows an inelastic case of collision, but the following reasoning applies to all types of collision (Figure 4.2).

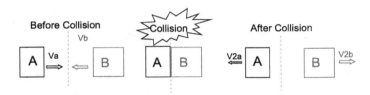

FIGURE 4.2
An elastic collision.

Before the collision, the Momenta of A and B are, respectively, denoted by $\overrightarrow{P_A^K}, \overrightarrow{P_B^K}$ in K and $\overrightarrow{P_A^{K'}}, \overrightarrow{P_B^{K'}}$ in K'.

After the collision, the objects A and B have different Momenta; their shapes may also change in case of inelastic collision, so let's change the names of these objects after collision: they become 2A and 2B. Subsequently, after the collision, their Momenta are in K: $\overrightarrow{P_{2A}^K}, \overrightarrow{P_{2B}^K}$ and in K': $\overrightarrow{P_{2A}^{K'}}, \overrightarrow{P_{2B}^{K'}}$.

In the frame K, the system Momentum conservation law states:

$$\overrightarrow{P_A^K} + \overrightarrow{P_B^K} = \overrightarrow{P_{2A}^K} + \overrightarrow{P_{2B}^K}. \qquad (4.5)$$

We need to show that equation 4.5 implies that we have in the frame K':
$$\overrightarrow{P_A^{K'}} + \overrightarrow{P_B^{K'}} = \overrightarrow{P_{2A}^{K'}} + \overrightarrow{P_{2B}^{K'}}.$$

Let's apply the Lorentz transformation to both sides of equation 4.5:

$$(\Lambda_w)(\overrightarrow{P_A^K} + \overrightarrow{P_B^K}) = (\Lambda_w)(\overrightarrow{P_{2A}^K} + \overrightarrow{P_{2B}^K}).$$

The Lorentz transformation being linear, we have:

$$(\Lambda_w)\overrightarrow{P_A^K} + (\Lambda_w)\overrightarrow{P_B^K} = (\Lambda_w)\overrightarrow{P_{2A}^K} + (\Lambda_w)\overrightarrow{P_{2B}^K}.$$

The Momentum being contravariant, we finally obtain: $\overrightarrow{P_A^{K'}} + \overrightarrow{P_B^{K'}} = \overrightarrow{P_{2A}^{K'}} + \overrightarrow{P_{2B}^{K'}}$ ∎

Let's write the spatial part of this relation:

$$m_A.\overrightarrow{V_A'}.\gamma_{V_A'} + m_B.\overrightarrow{V_B'}.\gamma_{V_B'} = m_{2A}.\overrightarrow{V_{2A}'}.\gamma_{V_{2A}'} + m_{2B}.\overrightarrow{V_{2B}'}.\gamma_{V_{2B}'}.$$

The velocities of the objects A, B, 2A and 2B in K' are obtained from those in K by applying the relativistic velocity composition law considering the velocity w of K' relative to K. We thus have:

$$V_{2A}' = -w \oplus V_{2A} = \frac{V_{2A} - w}{1 - wV_{2A}/c^2} \text{ and } V_{2A} = w \oplus V_{2A}' = \frac{V_{2A}' + w}{1 + wV_{2A}'/c^2}.$$

In particular, the problem encountered in the previous inelastic collision case §4.1.2 is solved as shown with Problem 5.5.

4.2.3 The Momentum Defined with Homogeneous Units

4.2.3.1 The Momentum with Homogeneous Units

We saw with the time-distance equivalence that time and distance can be expressed with a common unit, forming a homogeneous and natural system of unit. We will use the natural system of units as it brings more simplicity in physical expressions, and even facilitates their understanding. This requires adapting physical quantities; hence, we will see how to adapt the Momentum:

We saw that the time can be expressed with the distance unit, thanks to the universal time to distance conversion method: $t \rightarrow ct$. Using natural units any event M (t, x, y, z) becomes: M (ct, x, y, z). Likewise, $\overline{dM}$ = (cdt, dx, dy, dz). The Momentum (equation 4.3) expressed in natural units then becomes:

$$\overrightarrow{P^K(t)} = m\gamma_{v(t)}\frac{(cdt, dx, dy, dz)}{dt} = m\gamma_{v(t)}(c, \overrightarrow{v}(t)) \blacksquare \qquad (4.6)$$

Note that the γ factor remains unchanged with homogeneous units since it involves v/c.

4.2.3.2 The Momentum with Fully Homogeneous Units

The above homogeneous Momentum definition (equation 4.6) actually is not fully homogeneous, since the time in the denominator is still expressed with a time unit, whereas it is expressed with a distance unit (ct) in the numerator. Physically, this is not consistent, but it is done for the purpose of keeping the same dimension for the relativistic Momentum as the classical one, i.e., a mass multiplied by a distance and divided by a time.

If we express the Momentum with fully homogeneous units, we must express the denominator with the same common unit as in the numerator: ct. The Momentum expressed in fully homogeneous units, denoted by P*, is then:

$$\overrightarrow{P'}(t) = m\gamma_{v(t)} \frac{(cdt, dx, dy, dz)}{cdt} = m\gamma_{v(t)}(1, \frac{\overrightarrow{v(t)}}{c}) = m\gamma_{v(t)}(1, \overrightarrow{\beta}(t)) \blacksquare \qquad (4.7)$$

We notice that with fully homogeneous (natural) units, the natural unit for the Momentum is the mass (kg).

4.3 Revolutionary Consequences of the Relativistic Momentum

In classical physics, the mass was the measure of the inertia of an object, and was constant. We will see that these two fundamental characteristics are false in Relativity, and we will then examine the meaning of the mass in Relativity.

4.3.1 The Inertia of an Object Increases with its Speed

The figure below shows the comparison between the classical momentum, mv, and the spatial part of the relativistic Momentum, mvγ (Figure 4.3):

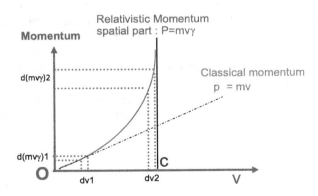

FIGURE 4.3
Comparison between the classical and the relativistic Momentum.

We can see that the relativistic Momentum curve starts out very close to the classical one (dotted), then becomes significantly higher and finally tends to the infinite as the object speed tends to c. This implies that for the same incremental speed increase, the relativistic Momentum increases much more rapidly when the object speed is close to c, than for usual speeds (as shown in the figure with dv1 and dv2).

We will further see that the Newtonian-like Force is: $\vec{F} = \dfrac{d\vec{P}}{dt}$. Hence, its spatial part, $\dfrac{d(m\vec{v}\gamma)}{dt}$, increases with the speed due to the presence of the γ factor, and tends to the infinite when the object speed tend to c.

Moreover, we will see in the next chapter that the energy required to accelerate an object increases with its speed, and tends to the infinite as the object speed tends to c.

This means that the inertia of an object, which is its resistance to its velocity change, is no longer constant, but increases with its speed, tending to the infinite as the speed tends to c. This is a very significant change regarding the concept of mass: the mass m no longer represents the inertia of an object, but the product mγ does; hence, this product **mγ was called "inertia"**.

One consequence is the impossibility for any object to reach the speed c, because the force required to accelerate this object would be infinite. However, particles having zero mass, such as the photon, do not face this impossibility (cf. more in § 5.4).

Remark 4.3: The product mγ was initially called "relativistic mass," but this name is progressively abandoned (cf. more in §5.2.5).

4.3.2 The Meaning of the Mass

4.3.2.1 The Mass is Not Necessarily Constant

In classical physics, the mass constancy resulted from experiments and observations, and was also supported by the system momentum conservation law. Indeed, the fact that $\vec{p} = \sum_{i=1}^{N} m_i \vec{v}_i = cst$ in all inertial frames mathematically implies that $\sum_{i=1}^{N} m_i = cst$, as shown in §4.4.1.

However, Relativity showed that the Momentum conservation means: $\sum_{i=1}^{N} m_i \gamma_i \vec{v}_i = cst$, but this relation does not imply the constancy of the sum of the masses. Hence some parts may incur mass changes, and indeed phenomena involving mass variations were observed but long after Relativity was discovered; more information is given in § 5.2.

Remark 4-4: In classical physics, the momentum constancy also implied the immobility of the center of gravity. In Relativity, this is not the case, as shown in Volume II §3.6.2.

4.3.2.2 The Lorentzian Norm of the Momentum is Invariant and Represents the Mass

We previously saw that the Momentum follows the Lorentz transformation when changing frame, meaning that it is a contravariant 4-vector. Besides, we saw that the Lorentz transformation leaves invariant the Lorentzian norm of such vector. Consequently, the Lorentzian norm of the Momentum, denoted by $\Delta s^2(\vec{P})$ and more usually by $\mathbf{P}^2$, is invariant, which is an important property.

The Momentum being generally expressed with homogeneous (natural) units[6], we first need to see how to compute the Lorentzian norm of a 4-vector expressed with homogeneous units:

Consider a random 4-Vector $\overrightarrow{OM}$ (t, x, y, z); its Lorentzian norm² is:

$$\Delta s^2\left(\overrightarrow{OM}\right) = c^2 t^2 - \left(x^2 + y^2 + z^2\right). \tag{4.8}$$

With homogeneous units, the time coordinate is expressed with the distance unit: $t_d = ct$. The same 4-vector $\overrightarrow{OM}$ expressed with natural units is then: (t_d, x, y, z). We can see that for $\Delta s^2(\overrightarrow{OM})$ to have the same value as equation 4.8, we must have: $\Delta s^2(\overrightarrow{OM}) = t_d^2 - x^2 - y^2 - z^2$.

According to relation 4.6, the Momentum expressed with homogeneous units is: $\vec{P} = m\gamma(c, \vec{v})$; its Lorentzian norm² is thus:

$$P^2 = \Delta s^2(\vec{P}) = m^2\gamma^2(c^2 - v^2) = m^2\gamma^2 c^2(1 - v^2/c^2) = m^2 c^2 \quad \blacksquare \tag{4.9}$$

We can see that the Momentum norm of an object represents its mass. As long as no force is applied on an object, its Momentum remains constant, which implies that its mass remains constant.

Note that the term c^2 stems from the fact that the Momentum is not expressed with fully homogeneous units. If we use the fully homogeneous Momentum definition (equation 4.7), then we have: $\mathbf{P}^{*2} = m^2$, showing that **physically the mass is the norm of the Momentum.**

Remark 4.5: Regarding the Lorentzian norm², we will prefer writing $\mathbf{P}^2$ rather than $ds^2(\vec{P})$ or $\Delta s^2(\vec{P})$; we will keep $ds^2()$ and $\Delta s^2()$ for space-time intervals. The bold character $\mathbf{P}$ is used to mean that P is a 4-Vector.

4.3.3 An Important Novelty: The System Mass

We saw that the Momentum of an isolated system, $\overrightarrow{P_{tot}} = \Sigma \vec{P_i}$, is constant. We will now show that its (Lorentzian) norm, $\Delta s^2(\overrightarrow{P_{tot}})$, is invariant. Let's then express the system Momentum in another frame K':

$$\overrightarrow{P_{Tot}} = \sum_{i=1}^{N} \vec{P_i} = \sum_{i=1}^{N} \Lambda(\overrightarrow{P_i}) = \Lambda\left[\sum_{i=1}^{N} \vec{P_i}\right] = \Lambda(\overrightarrow{P_{tot}}). \tag{4.10}$$

We used the Momentum contravariance and the Lorentz transformation linearity.

The relation 4.10 shows that the system Momentum is also contravariant, which implies that its Lorentzian norm is invariant. A notion which is both constant and invariant has a great importance in physics, as one of the main goals in science is to find invariants. Hence, a new concept was defined: the **system mass**, which is the norm of the system Momentum divided by c. The system mass thus is

$$M_{Tot}^2 c^2 = \mathbf{P}^2 = \left[\sum_{i=1}^{N} m_i c \gamma_i \right]^2 - \left[\sum_{i=1}^{N} m_i \vec{v_i} \gamma_i \right]^2 = cst \ \blacksquare \qquad (4.11)$$

Remark 4.6: This new system mass definition takes into account not only the masses of the different parts, but also their relative velocities. Previously, we mentioned that the sum of the masses of the parts of a system is not necessarily constant, which re-enforces the importance for this new system mass concept which is constant.

The physical reality of this system mass definition was confirmed by many experiments, and some will be presented in §5.2. Furthermore, this mass definition was also confirmed in General Relativity, as the relation giving the gravitational effects generated by a system uses this system mass definition.

Remark 4.7: The system mass is not the only system concept: for instance, heat also is a system property.

4.3.3.1 The Center of Momentum (CoM) Frame

The Center of Momentum (CoM) frame is defined as the inertial frame where the spatial part of the system Momentum is null, meaning: $\sum m_i \vec{v_i} \gamma_i = 0$.

From relation 4.11, we have: $M_{Tot}^2 c^2 = \mathbf{P}^2 = \left[\sum_{i=1}^{N} m_i c \gamma_i \right]^2$. $\qquad (4.12)$

Hence the system mass is: $M_{TOT} = \sum_{i=1}^{n} (m_i \gamma_i) \ \blacksquare \qquad (4.13)$

We saw that the system mass is invariant and constant for an isolated system. **Thus, the system mass is the sum of the inertias of all the parts in the CoM frame.**

One illustration is that if we heat a gas, its molecules will move faster, and consequently, the γ_i will increase and so the gas becomes heavier.

4.4 Complements

4.4.1 The Total Mass of an Isolated System Remains Constant in Classical Physics

We will show that in classical physics, the conservation of the sum of the masses of a system is a consequence of the closed system momentum conservation law.

Let's express the 3D system momentum conservation law in an inertial frame K before and after collision. The state before collision is denoted with the index 1, and that after collision is denoted with the index 2.

$$p_1^K = \sum_{i=1}^{N} m_{1i}^K v_{1i}^K = p_2^K = \sum_{i=1}^{N} m_{2i}^K v_{2i}^K$$

In a frame K' moving at the speed w relative to K, we have:

$$p_1^{K'} = \sum_{i=1}^{N} m_{1i}^{K'} v_{1i}^{K'} = p_2^{K'} = \sum_{i=1}^{N} m_{2i}^{K'} v_{2i}^{K'}.$$

Besides, we have: $v^{K'} = v^K - w$; hence:

$$p_1^{K'} = \sum_{i=1}^{N} m_{1i}^K (v_{1i}^K - w) = p_2^{K'} = \sum_{i=1}^{N} m_{2i}^K (v_{2i}^K - w).$$

Consequently: $p_1^{K'} = p_1^K - \sum_{i=1}^{N} m_{1i}^{K'} w = p_2^{K'} = p_2^K - \sum_{i=1}^{N} m_{2i}^{K'} w.$

Then, as $p_1^K = p_2^K$, we have: $\sum_{i=1}^{N} m_{1i}^{K'} w = \sum_{i=1}^{N} m_{2i}^{K'} w$, so: $\sum_{i=1}^{N} m_{1i}^{K'} = \sum_{i=1}^{N} m_{2i}^{K'}$ ∎

Conversely, Relativity showed that classical momentum conservation was false, which opens the possibility for the sum of the masses of an isolated system not to be constant.

4.4.2 Relativistic Momentum Demonstration in a 3D Approach Using an Inelastic Collision

Demonstrations of the Relativistic Momentum are generally more complex when using a 3D approach like in classical physics. The following one is an example and an exercise, where we will find the spatial part of the relativistic Momentum using a simple elastic collision scenario.

The classical momentum proved its validity for low speeds compared to c; hence, we are induced to assume that the relativistic Momentum has the form: $m.\vec{S}$, with $\vec{S}$ being a function of the 3D object velocity $\vec{V}$ (spatial part only). We do not know this function, but we will find it using the following scenario.

First, we will show that $\vec{S}$ must be extremely close to $\vec{V}.\gamma_V$. Second, we will show that this candidate solution, $\vec{S} = \vec{V}.\gamma_V$, satisfies the elastic collision case.

4.4.1.1 First Step: Demonstration That $\vec{S}$ Must Be Extremely Close to $\vec{V}.\gamma_V$

Consider an elastic collision between two objects having the same mass, for instance 2 billiard balls: one blue, one red. Initially, these balls are at opposite ends A and B, then we simultaneously send them together and with an equal impulse so that they move at the same speed and have a collision at the point C. After this collision, each ball is deflected from its original direction, and due to the symmetry of the scenario, the defection angle is the same for both balls. The ball A then goes to the point D such that AC = CD; the ball B to the point E such that CE = CB. Due to the symmetry, ADBE forms a rectangle (Figure 4.4).

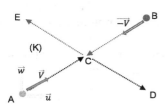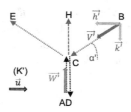

FIGURE 4.4
Elastic collision seen from two frames.

After leaving the point, the ball A velocity, denoted by $\vec{V}$, is constant until the collision. We denote by $\bar{u}$ its horizontal component along AD and $\bar{w}$ its transverse component along AE (which is perpendicular to AD). After the collision, the speeds (absolute value) are the same as before due to the kinetic energy conservation law, the collision being elastic, and the symmetry of the scenario.

Let's now consider this scenario from another inertial frame K' which is moving horizontally (along AD) at the speed $\bar{u}$ relative to the billiard frame, denoted by K. In K', the blue ball is seen moving vertically at the speed $\overline{W'}$ from A to C. Then this ball is seen moving downward and arrives at D, which is at the initial location of A in K'. Similarly, the red ball leaves the point B in K', goes at the speed $\overline{V'}$, hits the blue ball in C and then moves toward E.

4.4.1.1.1 Expression of the Momentum Conservation Law in K

The Momentum (spatial part) is assumed to be m. $\vec{S}$, with $\vec{S}$ being a function of $\vec{V}$(dx/dt, dy/dt, dz/dt), and we want to find this function; hence in the meantime, we will set:

$$\vec{S} = \vec{V}.\Psi(V). \tag{4.14}$$

When we consider the magnitude of a vector, we skip the arrow as we did with V and S.

We know that for usual speeds, meaning when V ≪ c, Ψ ≈ 1 since the classical momentum proved its validity for low speeds compared to c. Note that Ψ(V) depends only on V and not on its direction due to the universal isotropy. In the frame K, the Momentum conservation is respected for symmetry reasons. *Let's then examine the Momentum conservation law in K':*

4.4.1.1.2 Momentum Conservation Law in K'

First, the function Ψ must be the same in K and in K' due to the frame equivalence postulate. We will successively calculate the variation of the blue ball Momentum, and then the red one:

4.4.1.1.2.1 Momentum Variation of the Blue Ball in K'
From A to C, the Momentum of the blue ball is: $\overrightarrow{P^b}$ = m. $\vec{S'}$ = m.$\overline{W'}.\Psi(W')$. Then, from C to D, its speed is $-W'$, and so the variation of its Momentum is:

$$\overrightarrow{\Delta P^b} = -2m.\overrightarrow{W'}.\Psi(W'). \tag{4.15}$$

4.4.1.1.2.2 Momentum Variation of the Red Ball in K′   The red ball is going at the speed $\overrightarrow{V'}$ from B to C, with the horizontal component of $\overrightarrow{V'}$ being denoted by $\overrightarrow{h'}$ and the vertical one denoted by $\overrightarrow{k'}$. Since $\overrightarrow{h'}$ remains unchanged after the collision, it does not take part in the Momentum variation of the red ball. The variation of the Momentum of the red ball is then:

$$\overrightarrow{\Delta P'^{\tau}} = -2m.\overrightarrow{k'}.\Psi(V').$$ (4.16)

Let's calculate the speed k′ (which is the vertical component of V′)′: to this end, let's have another frame K″ which is moving at the speed −u relative to K. In K″, the scenario is the same as in K′, but symmetric: the red ball is moving vertically, and at the speed $\overrightarrow{W''}$ which is equal to $\overrightarrow{-W'}$ due to the symmetry relative to K (Figure 4.5).

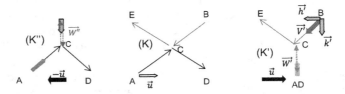

FIGURE 4.5
Elastic collision seen from K and two other symmetrical frames.

The speed $\overrightarrow{V'}$ can be considered as the result of the composition of the transverse speed $\overrightarrow{W''}$ in K″ with the speed of the frame K′ relative to K″, this speed being: u ⊕ u. Likewise, the speed $\overrightarrow{h'}$ is: −(u ⊕ u). Hence, the transverse component of the speed $\overrightarrow{V'}$, which is k′, can be calculated from W″: k′ = W″/$\gamma_{h'}$. Then the symmetry of the scenarios in K′ and K″ relative to K implies that W″ = W′. Hence:

k′ = W′/$\gamma_{h'}$ and subsequently: $\overrightarrow{\Delta P'^{\tau}} = 2m.\overrightarrow{W'}/\gamma_{h'}.\Psi(V')$. (4.17)

4.4.1.1.2.3 Expression of the Momentum Conservation in K′

From equations 4.15–4.17, we obtain:

$0 = \overrightarrow{\Delta P'^{\tau}} + \overrightarrow{\Delta P'^{b}} = 2m.W'/\gamma_{h'}.\Psi(V') - 2\,m.\overrightarrow{W'}.\Psi(W')$.

Hence: $\psi(V') = \psi(W').\gamma_{h'}$. (4.18)

4.4.1.1.3 Deduction of the Function ψ *by Examining Particular Cases*

If we consider the cases where the angle α′ between $\overrightarrow{V'}$ and the horizontal is very small, then we have: h′ = V′.cos α′ ≈ V′. Then equation 4.18 becomes: $\Psi(V') \approx \gamma_{V'}.\Psi(W')$.

If we consider the cases where W′ is very small compared to c, which is the case when the angle α′ is very small and the horizontal speeds are small compared to C, then we have:

$\Psi(W') \approx 1$, and then: $\Psi(V') \approx \gamma_{V'}$, meaning that S ≈ V. γ_V ∎

4.4.1.2 Second Step: Momentum Confirmation with an Elastic Collision

We will show that in the previous elastic collision, the 3D Momentum expression $\vec{P} = m\vec{V}\gamma_v$ satisfies the Momentum conservation law in K and K′ (Figure 4.6).

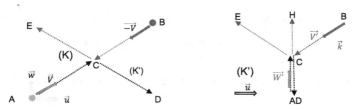

FIGURE 4.6
Elastic collision seen from two frames.

In the frame K, the new Momentum defined by this expression satisfies the Momentum conservation law before and after the collision, thanks to the symmetry of this scenario, with in particular the equality of the balls speeds before and after the collision. However, in the frame K′, this reasoning does not apply; hence, a demonstration is necessary:

4.4.1.2.1 Momentum Variation of the Blue Ball in K′

We will use the fact that $\vec{V} . \gamma_V$ is the spatial part of the proper velocity: $\dfrac{\overrightarrow{dM}}{d\tau}$

(cf. §4.2.2.1.1). In this context, we will consider the spatial parts of the 4D quantities, as if we didn't know they are 4D by essence.

The Momentum of the blue ball before collision is: $\overrightarrow{P_i^b} = m . \overrightarrow{S_i^b} = m . \dfrac{\overrightarrow{dM}}{d\tau^b}$ with $\overrightarrow{dM}$

representing the small portion of the ball trajectory that has been during $d\tau^b$.

The proper speed $\overrightarrow{S_i^b}$ is constant before the collision; hence, we have:

$\dfrac{\overrightarrow{dM}}{d\tau^b} = \dfrac{\overrightarrow{AC}}{\tau_{AC}^b}$ with τ_{AC}^b being the time taken by the blue ball to go from A to C

and measured with its proper time. Hence: $\overrightarrow{P_i^b} = m . \dfrac{\overrightarrow{AC}}{\tau_{AC}^b}$.

Similarly, the Momentum of the blue ball after collision is: $\overrightarrow{P_f^b} = m . \dfrac{\overrightarrow{CD}}{\tau_{CD}^b}$.

The Momentum variation of the blue ball then is:

$$\overrightarrow{P_f^b} - \overrightarrow{P_i^b} = m . \frac{\overrightarrow{CD}}{\tau_{CD}^b} - m . \frac{\overrightarrow{AC}}{\tau_{AC}^b} = 2m . \frac{\overrightarrow{CD}}{\tau_{AC}^b}. \tag{4.19}$$

We used the symmetry of the scenario before and after the collision.

4.4.1.2.2 *Momentum Variation in K' of the Red Ball*

The Momentum of the red ball before collision is: $\overrightarrow{P_i^r} = m.\dfrac{\overrightarrow{dM}}{d\tau^r}$ with $\overrightarrow{dM}$ being a small space-time interval along the trajectory of the red ball, and $d\tau^r$ the variation of the proper time of this red ball during $\overrightarrow{dM}$.

We then apply the same reasoning as with the blue ball, leading to: $\overrightarrow{P_i^r} = m.\dfrac{\overrightarrow{BC}}{\tau_{BC}^r}$ with τ_{BC}^r being the time taken by the red ball to go from A to C, and measured with the proper time of this red ball.

Similarly, the final Momentum of the red ball after collision is: $\overrightarrow{P_f^r} = m.\dfrac{\overrightarrow{CE}}{\tau_{CE}^r}$.

The Momentum variation of the red ball in K' is: $\overrightarrow{P_f^r} - \overrightarrow{P_i^r} = m.\dfrac{\overrightarrow{CE}}{\tau_{CE}^r} - m.\dfrac{\overrightarrow{BC}}{\tau_{BC}^r}$.

For symmetry reasons, we have: $\overrightarrow{CE} - \overrightarrow{BC} = 2\,\overrightarrow{CH}$, and we also have: $\tau_{CE}^r = \tau_{BC}^r$. Then the Momentum variation of the red ball in K' is:

$$\overrightarrow{P_f^r} - \overrightarrow{P_i^r} = 2m.\frac{\overrightarrow{CH}}{\tau_{BC}^r}. \tag{4.20}$$

We will now show that equation (4.19) is the opposite of equation (4.20), meaning that the system Momentum is conserved.

4.4.1.2.3 *The Momentum Definition, $m.V.\gamma_V$, Satisfies the System Conservation Law in K'*

First, we notice that $\overrightarrow{CH} = -\overrightarrow{CD}$ because of the invariance of the transverse distances between K and K'. In K, C is in the center of symmetry; hence, in K', the point C is in the middle of AH. Next, the time taken in K by the blue ball to go from A to C, and measured with its proper time, is the same as the time taken by the red ball for going from B to C, also measured with its proper time. Hence: $\tau_{AC}^b = \tau_{BC}^r$.

Besides, the proper time is invariant, meaning that in K' (as in any frame), we have: $\tau_{AC}^b = \tau_{BC}^r$.

> This can be illustrated as follows: let's place a perfect clock inside each ball. The value of τ_{AC}^b is the time duration measured by the blue clock from A to C, and seen by an observer who is co-located with this blue ball. It is clear that this value is independent of the frame from which this scenario is considered.

So we finally have: $2m.\dfrac{\overrightarrow{CD}}{\tau_{AC}^b} = -2m.\dfrac{\overrightarrow{CH}}{\tau_{BC}^r}$ ∎

4.5 Questions and Problems

Question 4.1: Is it necessary to express the Momentum with homogeneous (or natural) units?

Question 4.2: What is the difference between homogeneous (or natural) units and fully natural units?

Question 4.3: The temperature of a lake is 5°C in winter, and 27°C in summer. Is its system mass the same in summer and in winter?

Question 4.4: What is the Momentum of an object in its own frame?

Problems regarding classical physics and then Relativity

Problem 4.1: Combination of Inertial and Gravitational forces

You are in a lift and there is a weighing scale inside the cabin. You weigh yourself and you see that your weight has decreased.

Q1: What can you conclude regarding the lift motion?

Q2: You see that you are weighing 10% less than your normal weight. What can you conclude regarding the lift acceleration?

Problem 4.2: Calculation of the Altitude of a Geostationary Satellite. Relation between its speed and its altitude

Calculate the altitude that a satellite of mass m must have in order to be geostationary, meaning that any observer on Earth always sees this satellite in the same direction and at the same distance. Such satellites are very useful for broadcasting television or relaying telecommunication signals: the emitter and the antennas always point to the same direction. Does the satellite mass play a role? Please explain.

Problem 4.3: The Keplerian Orbital Law Relating the Velocity of a Planet with its Distance to the Star

Q1A: What is the relation between the speed V of a planet orbiting circularly around a star of mass M and the distance R of this planet to center of this star?

Q1B: Same question, but between the angular speed of the planet, ω, and R.

Q2A: The distance Mercury-Sun is 0.47 times the distance Earth-Sun (which is 1 Astronomical Unit). How many days does Mercury take to make one orbit around the Sun?

Q2B: Earth's speed in its orbit around the Sun is 107,000 km/h; what is Mercury's speed?

Problem 4.4: Weight is not only Gravity

A person weighs himself with a very precise weighing scale at a place somewhere on the equator, and sees: 70.000 kg. He does the same at the North Pole.

Q1: Is his weight greater, smaller or equal? (Earth radius is 6,371 km.)

Q2: By how much, if we assume gravity is $g = 9.81$ m s^{-2} in both cases?

Q3: Actually the Earth radius is larger at the equator than at the Pole by 22 km due to the centrifugal force; what impact does it have on your answer to Q2 (make a first-order approximation)?

Problems regarding Relativity

Problem 4.5: An Inelastic Collision

An object A of mass 2 kg goes at the speed V along the OX axis of an inertial frame K. Another object B of mass 1 kg goes at the speed 2V in the opposite direction. They will have an inelastic collision whereby both objects will be stuck together, forming an object C.

Q1: In classical physics, what is the speed of C in K?

Q2: Is it the same in Relativity? If not, in which direction will the object C go?

Q3: Will the mass of the object C be equal to the sum of the masses of A and B before collision?

Problem 4.6: The Contravariance Property 1

An object has a mass of 2 kg and is at rest in its rest frame K. Express its Momentum in K. A frame K' is moving at the speed 0.7 c relative to K along the OX axis.

Q1: Use the contravariance property to express the object Momentum in K'. We give gamma (0.7 c) $\approx$ 1.40.

Q2: Check that we obtain the same values when directly applying the Momentum definition in K'. **Q3:** Check that the Lorentzian norm of the Momentum is the same in K and in K'.

Problem 4.7: The Contravariance Property 2

An object with a mass of 2 kg is moving at the speed of 0.3 c along the OX axis of an inertial frame K.

Q1: Express its Momentum in K.

Q2: A frame K' is moving at the speed 0.7 c relative to K along the OX axis. Use the contravariance property to express the object Momentum in K'. We give: gamma (0.3 c) $\approx$ 1.048 and gamma (0.7 c) $\approx$ 1.40. Then check that the Lorentzian norm of the Momentum is the same in K and in K'.

Problem 4.8: The Momentum Density defined with the Mass Density

Consider a cube of mass m and volume V. It is at rest in the inertial frame $K°$, and its density in $K°$ is denoted by ρ_0. In any inertial frame K, we define the Momentum density vector as the 4-Vector $\vec{D} = \rho_0 . \vec{S}$, with $\vec{S}$ being the object proper speed, and ρ_0 its density in $K°$.

Q1: Is the vector $\vec{D}$ contravariant?

Q2: K is moving at the velocity $\vec{v}$ along the direction of one side of the cube. Express the components with homogeneous units of the 4-Vector $\vec{D}$ in the frame K.

Q3: The density of this cube considered in K is denoted by ρ. Express ρ as a function of ρ_0 and v.

Q4: Express the components of the 4-Vector $\vec{D}$ in K as a function of ρ.

Q5: Is the vector $\vec{E} = \rho.\vec{S}$ contravariant?

Notes

1 This principle (law) is also called "reciprocity of forces".
2 A collision is said to be elastic if the sum of the kinetic energies is conserved, meaning that no part of the kinetic energy has been exchanged with another sort of energy (thermal, electrical, etc.). Typical example: collision between billiard balls.
3 Cf. more in Volume II §3.3.5.
4 A demonstration based on an elastic collision scenario is in §4.4.1.
   • *A demonstration based on energy conservation is in Volume II § 3.1.2.*
   • *A mathematical demonstration using any type of collision is in Volume II §3.6.4.*
   • *A demonstration using the Lagrangian approach is in Volume II §3.5.2.*
   • *A physical demonstration based on an axiomatic analysis is in Volume II § 7.2.*
5 A scalar quantity characterizing an intrinsic notion is necessarily invariant (cf. Volume II §1.3.2).
6 Cf. §4.2.3.
7 The reader may wonder why k' is not equal to −W' since the vertical projection of the trajectory of the red ball between B and E is equal to the trajectory of the blue ball between A and C. The answer is that these balls don't take the same time to cover these distances: in particular, the red ball does not leave the point B in K' at the same time as the blue ball leaves A (due to the relativity of simultaneity).

5

The Energy Revolution: $E = mc^2\gamma$ and Its Consequences

Introduction

We saw that the relativistic Momentum has a time part, which raises the question of its meaning. We will see the reasons why it is the energy, and subsequently, we will carefully analyze the revolutionary consequences brought by this new expression: $E = mc^2\gamma$. The first novelty is the dramatic energy increase when the object speed is close to c. The second one, which is the most important consequence of Special Relativity as Einstein prophetically said, is the presence of a fixed part, mc^2. We will see that it expresses an equivalence of inertia and energy: mass can be transformed into energy and vice versa. Finally, physical consequences for the photon will be presented: we will see how the relativistic Momentum can be applied to the photon, explaining various phenomena including the famous photoelectric effect.

5.1 The Relativistic Energy Expression

5.1.1 The Relativistic Energy and the Time Part of the Momentum

In classical physics, the energy proved to be a fundamental concept, mainly due to the first principle of thermodynamics: the total energy of an isolated system remains constant. The amount of energy acquired by some of its parts is equal to the amount of energy lost by the other parts. We recall that:

- An isolated system (also called a closed system) exchanges no energy with its exterior.
- The total energy of a system is the sum of the energies of all its parts.

DOI: 10.1201/9781003201335-5

- The energy is transferable from one domain to another (e.g., from electromagnetic to mechanical or thermal or chemical, etc.), while respecting the constancy of the energy of the overall closed system.
- The kinetic energy of a moving object is ½ mv², and its variation is equal to the work of the force applied to this object (Coriolis law).

In classical physics, when considering the simple case of a collision between two objects, their trajectories after collision could be predicted, thanks to the two fundamental conservation laws: system momentum and total energy conservation. However, Relativity showed that the momentum definition was not correct, and that the relativistic Momentum has a fourth dimension. Hence, applying the system Momentum conservation law to the time part of relativistic Momentum provides an additional relation, and we will see that this relation relates to the energy.

Let's then consider the simple case of an elastic collision between two objects having collinear speeds:

Reminder: A collision is elastic if the sum of the kinetic energies of the objects before the collision is equal to the sum of the kinetic energies after the collision, meaning that no part of the initial kinetic energy has been transformed into other types of energy, such as heat, stress, or deformations (Figure 5.1).

In classical physics, the speeds of both objects after their collision could be calculated by solving the following system of two equations:

A. The system momentum remains constant:

$$m_A V_A + m_B V_B = m_A V_{2A} + m_B V_{2B} . \tag{5.1A}$$

B. The sum of the kinetic energies of the two objects also remains constant:

$$1/2 m_A V_A^2 + 1/2 m_B V_B^2 = 1/2 m_A V_{2A}^2 + 1/2 m_B V_{2B}^2 . \tag{5.1B}$$

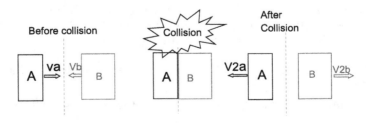

FIGURE 5.1
Elastic collision.

Let's now consider this scenario in Relativity: the relativistic Momentum still has this very conservation property for closed systems, and this law generates one additional equation compared to the classical momentum, due to the existence of the time part of the relativistic Momentum. This additional equation thus provides us with the following system of two equations which enables us to calculate the speeds of the two objects after collision:

Spatial part:

$$m_A V_A \gamma_{V_A} + m_B V_B \gamma_{V_B} = m_A V_{2A} \gamma_{V_{2A}} + m_B V_{2B} \gamma_{V_{2B}} . \tag{5.2A}$$

Time part, using the Momentum expressed with homogeneous units, cf. §4.2.3:

$$m_A c \gamma_{V_A} + m_B c \gamma_{V_B} = m_A c \gamma_{V_{2A}} + m_B c \gamma_{V_{2B}} . \tag{5.2B}$$

This implies that the time part of the Momentum of an object has a relation with its kinetic energy, since this system of equations enables us to calculate the trajectories of the objects after collision, which was previously done, thanks to equations 5.1A and 5.1B, the latter expressing the total kinetic energy conservation.

We are then induced to examine the time part of the Momentum, which we will denoted by P_t:

We have: $P_t = mc\gamma$. The γ factor is a function of the object speed, v, and for usual velocities, the ratio $v/c = \beta$ is extremely small so that γ can be approximated using the following Taylor polynomial expression (cf. Volume II Section 1.5.2):

$$\gamma(\beta) = 1 + \frac{1}{2}\beta^2 + \frac{3}{8}\beta^4 + R_4 , \tag{5.3}$$

this term R_4 is negligible compared to β^5.

For usual speeds, we even can make the approximation:

$$\gamma(v) = (1 - v^2/c^2)^{-1/2} \sim 1 + \frac{1}{2}v^2/c^2, \tag{5.4}$$

this approximation being better than β^3.

As an example, if the speed v is 100.000 km/h, which is extremely high for our usual activities, the approximation of $\gamma(v)$ given by equation 5.3 is in the range of 10^{-16}, and by equation 5.4 is in the range of 10^{-12}.

We then have: $P_t = mc \, \gamma \sim mc(1 + \frac{1}{2}v^2/c^2) = mc + \frac{1}{2}mv^2/c$.

We notice that the second term has a similarity with the classical kinetic energy $(1/2mv^2)$, so that if we multiply the time part of the Momentum by c, we obtain:

$$P_t c = mc^2\gamma \sim mc^2 + \frac{1}{2}mv^2 . \tag{5.5}$$

Remark 5.1: The multiplication of P_t by the speed c provides the product $P_t.c$ with the same energy dimension as in classical physics (i.e., the joule). Cf more in Sections 5.1.2.1 and 5.1.2.2.

The term $P_t.c$ is a good candidate for being the relativistic energy, despite the presence of the first term, mc^2, because:

- When making the difference between two energy states of an object, we obtain a result which is very close to the classical kinetic energy variation; indeed: $\Delta(P_t c) = \Delta(mc^2\gamma) \sim \Delta\left(\frac{1}{2}mv^2\right)$.
- In classical physics, it was always the energy variation between two states of an object or a system which mattered. Likewise, the mass was assumed to be always constant. Consequently, using equation 5.5 as the energy expression in the context of constant mass makes the term mc^2 disappear.

The expression

$$\mathbf{E} = \mathbf{P_t}\ \mathbf{c} = \mathbf{mc^2\gamma} \tag{5.6}$$

satisfies the validity conditions for being the relativistic energy definition because:

- It gives extremely close results to the classical law for usual speeds.
- It respects the fundamental principle of thermodynamics, which now is a consequence of the relativistic Momentum conservation principle (for isolated systems).
- All experiments confirmed this law; indeed, many have been made inside particle accelerators involving extremely high speeds such as 0.9999 c.
- Moreover, the law stating that the energy variation is equal to the work of the Newtonian-like Force is confirmed, as shown in Volume II Section 3.3.2.

Additionally, three other demonstrations of the famous energy relation 5.6 are presented:

- One showing that the conservation of the 3D quantity $\Sigma m_i \vec{v_i} \gamma_{v_i}$ for a closed system implies the existence of a scalar which matches with the energy (cf. Volume II Section 3.1.1).

- One assuming the law stating that the energy variation is equal the work of the force (cf. Volume II Section 3.6.5).

- One using the Lagrangian approach (cf. Volume II Section 3.5.2).

We will see in Section 5.2.2 the important consequences of the term mc², but before, let's clarify the reason why we had to multiply the term P_t by c, and for this we will first consider the energy with fully natural units.

5.1.2 The Energy with Fully Natural Units and with the International Unit (Joule)

5.1.2.1 The Energy with Fully Natural Units

We mentioned that physical expressions expressed with fully natural units are simpler and show their original and natural dimension. This is also true for the energy as we will see:

The energy is the time part of the Momentum, since the fundamental energy property, which is its conservation for an isolated system, stems from the Momentum conservation principle. We saw that the Momentum expressed with fully natural units, denoted by P*, is: $P* = (m\gamma, m\beta\gamma)$. The energy with fully natural units, denoted by E*, is then:

$$E* = P_t^* = m\gamma . \tag{5.7}$$

We can thus see that the natural dimension of the energy is the mass (kilogram) as for the Momentum. Note that it is also possible to express the mass with an energy unit, and this is common practice when working on particles.

5.1.2.2 The Term c² Is there Because the International System of Unit Is Not the Natural One

One can multiply any physical equation by any constant term; for instance, equation 5.7 can become: $k.E* = k.P_t^*$. However, if the term k has a physical dimension, such as a speed s, the equality: $s.E* = s.P_t^*$ must also be verified in all frames. We saw with the time-distance equivalence that the only speed which satisfies this condition is the speed of light. We can thus multiply both sides of any physical equation by c, and we can even repeat this multiplication. We thus multiply equation 5.7 twice by c, and we obtain:

$$c^2.E* = c^2.P_t^* = c^2 m\gamma . \tag{5.7a}$$

Note that with natural units, $c = 1$ without dimension; hence, equation 5.7a is identical to equation 5.7.

Before addressing the revolutionary consequences of the new energy expression, we will express some first remarks.

5.1.3 First Remarks on the Relativistic Energy Expression

Let's calculate the energy expression $E(v) = mc^2\gamma(v)$ using the Taylor polynomial (equation 5.3), and we obtain:

$$E = mc^2\gamma(v) = mc^2 + \frac{1}{2}mv^2 + \frac{3}{8}mv^4/c^2 + \cdots \qquad (5.8)$$

The energy is thus composed of two parts: mc^2 and a series of terms which are a function of the object speed v; this series is the kinetic part of the energy.

5.1.3.1 First Remarks on the Term mc²

The term mc^2 is independent of the frame where the object is considered: it is the intrinsic part of the energy, and corresponds to the energy of the object at rest: this is an important novelty introduced by Relativity which will be further discussed. We notice that this intrinsic energy is huge due to the term c^2, meaning that our usual objects possess an enormous amount of energy. However, we do not notice it, the reason being that their masses are constant (or almost).

This intrinsic term provides an absolute scale to the energy. It is worth noting that it is not the first time in history that a physical quantity acquired an absolute scale: this was indeed the case of the temperature as W. Kelvin discovered in 1848 that there existed an absolute zero: –273°C. In temperature also, what only mattered was the difference between two states of a system.

5.1.3.2 First Remarks on the Kinetic Part

The relativistic kinetic energy is very well approximated by ½ mv^2 for low speeds compared to c.

> If, for example, the object speed v is below 2 million km/h, the difference between the relativistic kinetic energy and the classical one is below 0.1%. If the object speed is 108 million km/h, this difference is 5%. Beyond this speed, the difference rapidly increases.

The kinetic energy being frame dependent, the total energy is NOT an intrinsic notion.

5.1.3.3 A Useful Relation: $E^2 = m^2 c^4 + p^2 c^2$

There is a useful relation giving the split between the intrinsic and the kinetic parts of the energy; it is:

$$E^2 = m^2 c^4 + p^2 c^2. \tag{5.9}$$

where p is the 3D spatial part of the Momentum: $\vec{p} = m\vec{v}\gamma$.

Demonstration: From equation 4.9), we have: $\mathbf{P}^2 = P_t^2 - p^2 = m^2 c^2$.

Then, from equation 5.6): $P_t = E/c$; hence: $E^2/c^2 - p^2 = m^2 c^2$, so finally: $E^2 = m^2 c^4 + p^2 c^2$ ■

5.2 The Energy Inertia Equivalence and Other Revolutionary Consequences

We will first address the new kinetic energy and see an important result: if we want an object to reach the speed c, we need an infinite amount of energy. We will then see the famous equivalence between inertia and energy.

5.2.1 The Energy Required to Accelerate an Object to the Speed C Is Infinite

Let's compare the relativistic kinetic energy with the classical one: we saw that the relativistic kinetic energy is: $K = mc^2 (\gamma_v - 1)$.

We can see that when v tends to c, γ_v tends to the infinite, and so does the relativistic kinetic energy. This was not the case of the classical kinetic energy ($\frac{1}{2} mv^2$).

We also observed that for low speeds compared to c, the kinetic energy can be approximated by:

$$K = E - mc^2 \approx \frac{1}{2}mv^2 + \frac{3}{8}mv^4/c^2.$$

This shows that the relativistic kinetic energy is always greater than the classical one. The graphic below shows the comparison between the relativistic kinetic energy K and the classical one (Figure 5.2):

We can see that the energy required for an object to reach any speed is higher than in classical physics (the relativistic slope being beyond the classical one), and this difference increases up to the infinite as the object speed

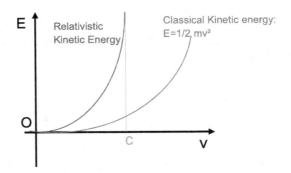

FIGURE 5.2
Comparison between the relativistic and the classical kinetic energies.

tends to c. This implies that we need an infinite energy to reach the speed c, meaning that it is impossible for any object to reach c, with the exception of massless objects such as the photon.

Note that this result is consistent with the one seen with the relativistic Momentum (cf. Section 4.3.3).

5.2.2 The Energy Inertia Equivalence and Nuclear Reactions

The relation $E = mc^2\gamma$ expresses an equivalence between the inertia of an object, $m\gamma$, and its energy. Moreover, if we use fully natural units, we have a perfect identity between the energy and the inertia[1], both being the time part of the Momentum ($m\gamma$). Thus, if an isolated object loses (resp. gains) an amount of inertia, $d(m\gamma)$, it simultaneously emits (resp. absorbs) an equivalent amount of energy: $dE = c^2.d(m\gamma)$.

The famous relation $E = mc^2$ gives the energy in the object's own frame (also called rest frame, so that $\gamma = 1$). In this frame, we can say after Einstein: "The mass is the measure of the energy contained in an object". Given the huge value of c^2, which is 8.99×10^{16} m²/s², a very small mass variation requires (or generates) a huge amount of energy: for example, if an object loses 1 g, it generates an energy of 10^{14} joules, which is in the range of the first atomic bomb. This explains why in our everyday life, no mass change was ever detected. However, Einstein prophetically said that with the progress of our instruments, we will be able to observe that a piece of radium continuously loses some mass due to its continuous emission of radiation.

5.2.2.1 Some Remarkable Cases of Exchanges Between Mass and Energy

Probably the most remarkable exchange between mass and energy is the thermonuclear fusion that occurs in the core of the Sun, and which is responsible for most of our energy. This phenomenon will be further detailed, but there are also other amazing phenomena involving exchange of mass, energy and radiation, such as:

- When an electron meets with a positron, both are annihilated and transformed into two very high-energy photons (gamma rays), the energy of each gamma ray corresponding to μc^2 (with μ being the electron's mass).
- Conversely, the creation of an electron-positron[2] pair is possible from the energy of gamma rays.
- Creation of a particle-antiparticle pair during inelastic reactions involving extremely high-speed particles, whereby part of their kinetic energies is transformed into mass.
- Energy can be transformed into matter (protons), but this requires an enormous concentration of energy. Besides, so far as we know there is no mechanism to produce matter without producing antimatter.

5.2.2.2 Mass Changes During Inelastic Collisions

We saw that the system energy conservation law concerns both elastic and inelastic collisions. In an inelastic collision, part of the kinetic energy is transformed into other sorts of energy: thermal, potential, binding energy, even matter, etc.

Consider an inelastic collision between two objects A and B, which is such that after the collision the two objects are stuck together and stay still in the frame K where they are observed (Figure 5.3).

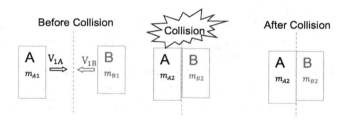

FIGURE 5.3
An inelastic collision.

The time part of the system Momentum conservation yields: $m_{a1}.c.\gamma_{a1} + m_{b1}.c.\gamma_{b1} = m_{a2}.c + m_{b2}.c$. We indeed took into account the fact that the speeds of A and B are null in the frame K after their collisions.

This relation shows that the masses of A and B are greater after their collision than before: they contain the kinetic energy that the objects had before the collision, and which was transformed into other sorts of energies (thermal, binding, etc.)[3].

Concerning phenomena of our day-to-day life, the values of γ are extremely close to 1; hence, these masses' increases are infinitesimal, but they exist. Conversely, for "ultra-relativistic" particles such as electrons in particle accelerators, the masses' increases are significant.

5.2.3 Nuclear Reactions: Fission and Fusion

Consider a nucleus AB, composed of two smaller nuclei, A and B. Suddenly, AB spontaneously incurs fission into A and B, and this fission triggers the rapid move of these nuclei in opposite directions to each other (Figure 5.4).

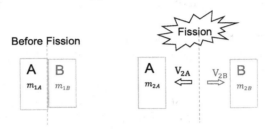

FIGURE 5.4
Fission.

The mass of the nucleus AB before fission is $m_{1A} + m_{1B}$. After fission, the masses of the smaller nuclei are, respectively, m_{2A} and m_{2B}. The energy conservation law yields:

$$m_{1a}.c + m_{1b}.c = m_{2a}.c.\gamma_{2a} + m_{2b}.c.\gamma_{2b}.$$

This relation shows that the masses of the nuclei A and B after fission are less than before: part of their masses within AB has been converted into kinetic energy during the fission.

Inversely, this shows that an atom whose mass is less than the sum of the masses of its parts when they are separate, cannot spontaneously incur fission into these parts; hence it is stable.

The smallest components of a nucleus are the **nucleons**: protons, neutrons. A nucleus doesn't spontaneously incur fission into all its nucleons; hence its mass, M, is less than the sum of the masses of its nucleons: $M < \sum m_i$. This **mass defect** corresponds to the **binding energy** of the nucleus, classically denoted by B:

$$B = \sum m_i c^2 - Mc^2 \quad \blacksquare \qquad (5.10).$$

The greater the binding energy B, the greater the mass defect, the greater the energy that was required to make the nucleus from its separate nucleons, the greater the external energy is required to trigger its fission; hence, the more stable the nucleus is.

Consider a nucleus whose binding energy is B: we mentioned that it cannot spontaneously fission into its nucleons, but it can into nuclei provided that the sum of their binding energies is greater than B. Indeed, this fission leads to an overall increase of the system binding energy (i.e., the sum of the binding energies of the system), meaning an increase of the overall mass defect, which corresponds to the energy that is converted during the fission into kinetic energy and also radiation and thermal energy (which is a form of kinetic energy).

A nucleus is characterized by its mass number, classically denoted by A, which is the sum of the number of its protons, Z, and the number of its neutrons: $A - Z$. The ratio B/A is an important indicator to determine if fission is possible: indeed, if each resulting nucleus has a greater B/A ratio than the initial nucleus, then fission is spontaneously possible as it leads to a greater overall mass defect. (The corresponding demonstration is given with the problem 5.17).

This fission condition is met by some atoms that are heavier than Iron (Fe) in the Mendeleev table. These can indeed disintegrate into smaller atoms that have a higher B/A ratio, as shown in the Figure 5.5.

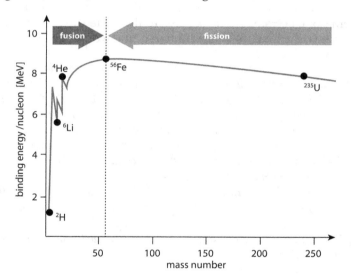

FIGURE 5.5
Diagram showing the evolution of the B/A ratio as a function of A (Image from Shutterstock.)

A famous example of fission is the heavy atom of uranium-235. It spontaneously incurs fissions into smaller atoms while generating a high amount energy through radiation and emission of neutrons. However, such spontaneous fission is very rare (U-235 has a half-life of millions of years), but a collision with a neutron can create U-236, which is much more unstable.

Fusion: The same reasoning shows that a spontaneous fusion between nuclei is possible if it leads to a greater nucleus having a greater B/A ratio than the initial ones. Indeed, the overall mass defect then increases, being transformed into radiation, kinetic and thermal energy. This fusion condition is met by some atoms that are lighter than the iron in the Mendeleev table.

A typical example is the thermonuclear fusion that synthesizes helium in the core of the Sun. There are several processes that make such a reaction, the most common one being the "proton-proton chain[4]" which comprises several steps and can be summarized as follows:

4 protons →1 helium-4+2 neutrinos+gamma rays. An amount of 0.7% of the initial masses is converted into radiation (cf. more in Volume II Section 3.2.2).

The energy generated by a nuclear reaction is much greater than by any chemical reaction because it involves the nuclear strong force, which links the nucleons together. Conversely, chemical reactions involve the electromagnetic force, which is much weaker than the Strong force. Most of the nucleon mass is due to the energy field of the strong force. If we further look inside the nucleons (protons and neutrons), we find quarks which also incur important interactions representing an energy and thus a mass. Besides, part of the mass of the atom is also due to the binding energy of the electrons, this energy being of electromagnetic nature. To illustrate the combined effect of all these actions, the total mass of a human is made of roughly 90% of energy and 10% of rest mass.

Remark 5.2: *At the time Relativity was found, all these developments were not conceivable. Several important steps were indeed necessary: the Neutron was discovered in 1932 by J. Chadwick (team of Rutherford). In 1938, the German scientist F. Strassmann (team of O. Hann and L. Meitner) used the Neutron as a bullet toward an atom of Uranium-235, and performed its fission into two atoms of approximately half size, but with a mass loss. This reaction generated much more energy than in usual chemical reactions, which was consistent with Einstein's famous relation when taking into account the mass losses observed. The next year, the French scientist Joliot-Curie showed the theoretical possibility of a chain reaction, whereby the fission of an atom of Uranium-235 could generate more neutrons that would trigger the fission of other Uranium-235 atoms, opening the way to the atomic bomb. However, huge practical problems still remained. (The USA needed more than 5 years and 130,000 men to achieve it.)*

5.2.4 Units Adapted to Particles and Typical Nucleons Masses

At the particle level, the commonly used unit for the energy is the MeV ($=10^6$ eV) or GeV ($=10^9$ eV).

One eV is the kinetic energy acquired by one electron evolving in an electrical potential difference of 1 Volt. Thus: 1eV$=1.602\ 176\ 634 \times 10^{-19}$ joule. (Likewise, 1 Volt = 1 Joule per Coulomb).

For the Momentum (impulse), the unit is: MeV/c or GeV/c; and for the mass: MeV/c² or GeV/c².

Alternatively, the atomic mass unit (abbreviation u) has been defined so that the mass of the most abundant isotope of carbon-12 is 12u. Hence: 1u$=1.66 \times 10^{-27}$ kg$=0.9315$ GeV/c².

We then have:
Proton mass: 938.3 MeV/c² = 1.000728 u = 1.673×10^{-27} kg.
Neutron mass: 939.6 MeV/c² = 1.00867 u = 1.675×10^{-27} kg.
Electron mass: 0.511 MeV/c² = 5.49×10^{-4} u = 9.093×10^{-31} kg.

Thus, the neutron and the proton masses are close to: $1\ \text{GeV}/c^2 = 1.783 \times 10^{-27}\,\text{kg}$.

Finally, fully natural units are also used whereby c and v, respectively, become 1 and β. For example, the equation 5.9 becomes: $E^2 = m^2 + p^2$. The eV then becomes a common unit for the energy, the mass and the Momentum.

5.2.5 Naming and Physical Considerations

The terminology "equivalence between energy and *inertia*" is preferable to "equivalence between energy and *mass*" because it is the inertia, and not the mass, which is equivalent to the energy. It is only in the object's own frame that its mass is equivalent to its energy: it is its intrinsic energy, and it is the minimal energy that was required to create the object; in any other frame, the energy has an additional kinetic part.

To illustrate the confusion generated by the expression "mass-energy equivalence", let's consider the case of the photon: it has an energy, but no mass; how can there be an equivalence between the two?

This inadequate nomenclature stems from another inadequate expression: the "relativistic mass". Initially, when Relativity showed that the inertia of an object increases with its speed, the product $m\gamma$ was called the "relativistic mass", and the mass m was called the "rest mass". However, this relativistic mass concept has been progressively abandoned, since there is another concept related to the mass and which is more fundamental: the mass defined as the Lorentzian norm of the Momentum (divided by c). This new definition is applicable to an object as well as a system (the system mass), and has two fundamental properties: it is invariant (intrinsic) and constant when the object or the system is isolated.

Einstein introduced this term "relativistic mass", but he later recognized that this naming was not appropriate. Moreover, we saw in the relativistic Momentum demonstration that the γ factor appears because the classical speed, v, was not acceptable in Relativity and needed to be replaced with the Four-Speed that includes the γ factor. Consequently, it is not logical to assign the γ factor to the mass by calling $m\gamma$ the relativistic mass.

The Momentum is the cornerstone of the dynamics as it carries the energy, which means the capacity of change[5]. Consequently, the Momentum has also been called the "momenergy", and extended to the "energy-momentum tensor".

Fundamentally, it is worth noting that the word "equivalence" used in the context of "equivalence between energy and inertia" has a different meaning than in the "equivalence between time and distance": in the energy-inertia equivalence, equivalence actually means identity: these two names refer to two different properties of the same physical quantity, which is the time part of the Momentum. Indeed, with fully natural units, this quantity, $m\gamma$, is both the inertia and the energy. Conversely, in the case of the time-distance equivalence, time and distance are not identical and remain two different concepts (cf. Section 3.4).

Another amazing aspect of the energy-inertia equivalence is that it is an oxymoron, the concept of energy being the opposite of the concept of inertia: the capacity to make changes versus the capacity to oppose changes. A positive aspect of this opposition is that it prevents objects from surpassing the speed c, which would contradict the causality principle.

5.3 The Force in Relativity

Having seen that we are in a 4D space-time universe, the classical 3D Newtonian force must also be derived from a 4D Force. In classical physics, we saw that the 3D Newtonian force was derived from the classical momentum, $\vec{P} = m.\vec{v}$, with $\vec{f} = \dfrac{d\vec{p}}{dt}$, since the system momentum conservation principle explained two properties of the classical force: the reciprocity of forces, and the fact that the force only influences the second derivatives of the coordinates of an object, and not further.

Hence, we were induced to consider the relativistic Force as the time rate of change of the relativistic Momentum. However, Relativity brought major changes: notably, the Momentum has a time dimension which represents the object energy. Still, time is not absolute: there are many possible time sources, but the time must come from an inertial frame because the force is by essence what makes the object deviate from an inertial trajectory. Hence, there are two main possibilities for the time source of the relativistic Force:

- The first one is the time t of the inertial frame K where the Force is considered (meaning by an observer who is fixed in K). This leads to:

$$\vec{F} = \frac{d\vec{P}}{dt} = \left(\frac{d(mc\gamma)}{dt}, \frac{d(m\vec{v}\gamma)}{dt} \right), \qquad (5.11)$$

This Force is called the **Newtonian-like Force**, since it uses the same time t as the classical Newtonian force.

- The second one is the special case where the Force is considered in the inertial tangent frame[6] (ITF) of the object that incurs this Force. We saw that during a small time interval, the time of this ITF is identical to the object' proper time, denoted by τ. Hence, this Force, named the **Force of Minkowski** or the **Four-Force**, was defined as:

$$\vec{\Phi} = \frac{d\vec{P}}{d\tau}. \qquad (5.12)$$

From these definitions, we can derive the following properties:

1. There is a relation between the two relativistic Forces:

$$\vec{\Phi} = \frac{d\vec{P}}{d\tau} = \frac{d\vec{P}}{dt}\frac{dt}{d\tau} = \vec{F}.\gamma_v \ . \tag{5.13}$$

2. When $v \ll c$, the two relativistic Forces are very close (since $\gamma_v \approx 1$). Moreover, their spatial parts are very close to the classical Newtonian force: $f = \dfrac{d(m\vec{v})}{dt}$.

3. The time part of the Newtonian-like Force is the Power. Indeed, the time part of the Momentum is E/c; hence: $F_t = \dfrac{dE}{cdt}$.

4. The spatial part of the Newtonian-like Force is, assuming that the mass is constant:

$$\vec{F_s} = \frac{d(m\gamma\vec{v})}{dt} = m\gamma\vec{a} + \frac{d(m\gamma)}{dt}\vec{v} = m\gamma[\vec{a} + \vec{v}\gamma^2(\vec{\beta}\cdot\vec{\beta}')] \tag{5.14}$$

We used the derivative[7]: $\dfrac{d\gamma}{dt} = \gamma^3 \ (\vec{\beta}\cdot\vec{\beta}')$.

This relation shows that the Newtonian-like Force is not the same in all inertial frames (being dependent on v) and that it is not parallel to the acceleration, unless the Force and the velocity are parallel.

5. The Force of Minkowski is contravariant since $\vec{P}$ is contravariant and τ is invariant. Indeed, when changing frame from K to K', the expression of $\vec{\Phi}'$ can be obtained from ϕ by applying the Lorentz transformation:

$$\vec{\Phi}' = \frac{d\vec{P'}}{d\tau} = \frac{d(\Lambda)(\vec{P})}{d\tau} = (\Lambda)\frac{(d\vec{P})}{d\tau} = (\Lambda)\ \vec{\Phi} \ . \tag{5.15}$$

Remark 5.3: The Newtonian-like Force is not contravariant due to the term dt which refers to the observer's frame. Hence, it is not an intrinsic notion, unlike the Four-Force.

The properties of these relativistic Forces are further developed in Volume II.

**

We will now present physical considerations on the photon since it plays a key role in Relativity.

5.4 Physical Considerations on the Photon

The light speed invariance was the compelling event leading to Relativity. The same year 1905, Einstein explained the photoelectric effect by stating that the photon is the quantum of electromagnetic energy, light being electromagnetic waves. Moreover, he stated that the photon has both a wave and a corpuscle nature. These assertions were even more revolutionary than Relativity, and required an even longer period of time (15 years) to be accepted by the scientific community. However, the latter finally awarded Einstein the Nobel Prize in 1921 for his explanation of the photoelectric effect, which opened the way to quantum mechanics.

In a first part, we will present general considerations on the photon, then the photoelectric effect.

5.4.1 General Considerations on the Photon

5.4.1.1 The Photon Energy

In classical physics, any particle with zero mass had zero energy whatever its speed, its energy being $\frac{1}{2} mv^2$. Conversely, the new relativistic energy expression $E = mc^2\gamma$ allows massless particles to have an energy which is not null provided that they move at the velocity c: indeed, $mc^2\gamma$ is undetermined in such case (γ being infinite). Consequently, the photon, being a massless particle moving at the speed c, can now have an energy, and Einstein showed that it is:

$$E = h\,\nu \qquad\qquad (5.16)$$

where h is the Planck constant: $h = 6.62 \times 10^{-34}$ joule.second and ν the photon frequency.

Some history on the Planck constant: In the late 1800s, the spectrum of electromagnetic radiation (including light) of the black body raised an open issue: it was found that any matter emits photons with frequencies that only depends on the temperature of that matter[8]. At each temperature corresponds a frequency spectrum of the photons emitted, and all these spectra have the same shape as shown in the picture below. The issue was that this shape reflected two different laws, one for the high frequencies and another for the low frequencies (Figure 5.6).

Max Planck found in 1901 the mathematical equation matching the whole spectrum: his formula assumed that the energy of the oscillators within the black body was quantized, i.e., being in the form N·h·ν, where N is an integer. However, Planck believed that this quantization was a computational artifice and not a physical phenomenon, and consequently opposed Einstein's revolutionary statement that the photon was the quantum of electromagnetic energy.

Photons exist with an extremely wide range of energies, as shown in Table 5.1.

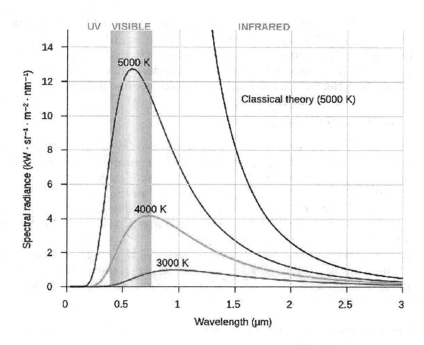

FIGURE 5.6

Black body spectrum. (Image from Shutterstock.)

TABLE 5.1

The Wide Range of Photons

Category	Wavelength	Frequency	Energy
Hard gamma	1×10^{-9} nm	3×10^{26} Hz	1.2×10^{12} eV
Gamma	1×10^{-6} nm	3×10^{23} Hz	1.2 GeV
Gamma/X-ray	0.001 nm	3×10^{19} Hz	12 MeV
X-ray	1 nm	3×10^{17} Hz	120 keV
X-ray/ultraviolet	10 nm	3×10^{16} Hz	12 keV
Ultraviolet	100 nm	3×10^{15} Hz	1.2 keV
Visible (blue)	400 nm	7.5×10^{14} Hz	3.1 eV
Visible (red)	700 nm	4.3×10^{14} Hz	1.8 eV
Infrared	10,000 nm	3×10^{13} Hz	0.12 eV
Microwave	1 cm	30 GHz	1.2×10^{-4} eV
Microwave/radio	10 cm	3 GHz	1.2×10^{-5} eV
Radio	100 m	3 MHz	1.2×10^{-8} eV
Radio	100 km	3 kHz	1.2×10^{-11} eV

Photons that are in the range of visible light have a very small energy, in the range of 2 eV, which is not quite sufficient to be individually detected by a human eye. Their extremely small energy explains why photons can be extremely numerous.

We saw that the photon moves at the same speed c in all inertial frames; hence, there is no inertial frame where the photon is at rest. On the other hand, Einstein stated that the photon also has a particle aspect, which appears when there is an interaction, such as a collision with an electron as it is the case with the photoelectric effect and the Compton effect (cf. Volume II Section 3.6.1).

5.4.1.2 The Photon Momentum

According to the relativistic Momentum definition, $(mc\gamma, mv\gamma)$, the spatial part of the photon Momentum equals the time part since the photon speed v equals c. This implies that the mass of the photon is null since the mass is the (Lorentzian) norm of the Momentum (with natural units).

Remark 5.4: From an axiomatic perspective, this suggests that the Momentum is a more primitive notion than the mass (cf. more in Volume II Section 7.2).

The spatial part of the photon Momentum, denoted by p, is equal to its time part, which equals E/c. Having seen that $E = h\,v/c$, the spatial part is: $E = h\,v/c$. The photon thus has a very small spatial momentum, but it is enough to produce effects at different scales:

- When an atom emits a photon, it incurs a small recoil. (This case will be further studied.)
- Some effects are visible, such as the tails of the comets: the pressure of radiation from the Sun is responsible for pushing away the dusty cloud made of very small particles around the comets. They form a tail in the opposite direction from the Sun. (These particles having a high surface/weight ratio, the radiation pressure offsets the gravitational forces which are unimportant around most comets.)
- Some effects are even huge: In case of extremely dense and energetic radiation such as in the kernel of the Sun, these photons exert a huge pressure that prevents the surrounding masses from falling to the center due to gravity.

The photon Momentum thus is:

$$\mathbf{P} = (hv/c, hv/c) . \tag{5.17}$$

We will see several cases of interactions between the photon and matter, but before that, an amusing scenario is presented, which illustrates the novelty of the relativistic mass:

5.4.1.3 An Atom that Emits a Photon Loses Mass

When an atom emits a photon, the system composed of the atom and the photon can be considered isolated; hence, its Momentum remains constant.

Before the photon emission, the atom Momentum in its rest frame is (Mc, 0) with M being the atom mass After collision, it loses the infinitesimal mass dm, and moves at the speed v. The constancy of the system Momentum yields:

$$(Mc, 0) = ((M - dm)\gamma c, (M - dm)v\gamma)) + (hv/c, hv/c)$$

We can see that the atom acquires a spatial momentum, meaning that it is pushed in the opposite direction from the photon: it incurs a recoil with the speed given by: $v\gamma = -hv/c(M - dm)$. As γ is very close to 1 and dm to zero, we have:

$$v \approx -hv/cM. \tag{5.18}$$

The reverse is also true: when an atom absorbs a photon, its mass increases and it is pushed in the photon direction.

Regarding the mass loss, dm, we have from the time components: $Mc = (M - dm)\gamma c + hv/c$. Hence, taking into account $v \ll c$:

$$dm = hv/c^2\gamma + M(\gamma - 1)\gamma \approx hv/c^2 \ \blacksquare \tag{5.19}$$

We thus have this ironic situation where the photon, which has no mass, actually conveys mass from its emitting body to its absorbing body. This case inspired Einstein with a thought experiment, the "photon box", presented in Volume II Section 3.6.2.

5.4.2 The Famous Photoelectric Effect

The French scientist A. Becquerel noticed in 1839 that when some metals were submitted to light, electrons were ejected. It was strange that the essential factor triggering the electron ejection was the frequency (color) of the light beam: below a certain frequency, no electron was emitted; beyond this frequency, electrons were ejected. Thus, there is a threshold in the light frequencies that triggers the electrons ejection (Figure 5.7).

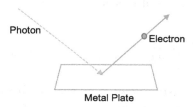

Photon

Electron

Metal Plate

FIGURE 5.7
Photoelectric effect.

Moreover, the kinetic energy of the ejected electrons was a linear (affine) function of the light frequency.

Conversely, the light intensity did not play any role in the electron emissions, nor in their kinetic energies after emission. This was disconcerting because light was thought to be a wave; hence, the energy of a light beam was thought to be proportional to its intensity independently of its frequency.

Moreover, the ejection of the photon was quasi-instantaneous at the time of the light absorption, which promoted the idea of a particle effect (instead of the accumulation of energy from a continuous wave).

Einstein explained in 1905 that the photon is the minimum quantum of electromagnetic energy, and that it has both a wave and a corpuscle aspect. Hence, when considering its corpuscle aspect, we can speak of a real collision between a photon and an electron; then, if the photon energy is beyond a certain threshold, it can eject the electron from its atom. Einstein stated that the photon energy follows the relation: $E = h.\nu$ (as previously mentioned), and that the threshold frequency corresponds to the electron binding energy in its atom, denoted by K_0. The minimum frequency required for a photon to eject the electron is then: $K_0 = h\nu_0$.

Correspondingly, a photon having a higher energy than K_0 communicates its remaining energy to the electron in the form of kinetic energy whose value is: $K = h\nu - h\nu_0 = h(\nu - \nu_0)$, which was confirmed by experiments.

*

The photoelectric effect has many applications, from solar cells to our vision, the photoelectric effect being at work in our retina cells.

5.5 Questions and Problems

Question 5.1: The Mass of a Stable Atom and of its Components
A stable atom A is made of two components B and C. Compare the mass m of the atom A with the sum of the masses of the components B and C.

Question 5.2: Natural units
Is the MeV a fully natural unit of mass? Is the kg a fully natural unit of energy?

Question 5.3: What are the Newtonian-like Force and the Four-Force in the object's own frame, assuming its mass is constant?

Problem 5.1: Natural units

An atom weighs 105 GeV, as expressed with fully natural units; what is its mass in Kg?

Problem 5.2: The Deuteron

The deuteron (also called heavy hydrogen) is made of one proton and one neutron. Its rest mass is: 1.876 GeV/c². Is it stable? What percentage of the mass has changed? What energy is required to break the deuteron into its nucleons?

Problem 5.3: Hydrogen Ionization

The hydrogen atom comprises one proton and one electron. It is possible to pull away the electron, meaning to ionize the atom, but this requires an energy of 13.58 eV. The hydrogen atom mass is 1.00783 u before ionization; does it change after ionization, and if yes, by what percentage?

Problem 5.4: The Mass of Two Photons Going in Opposite Directions is NOT Null

What is the system mass of an isolated system composed of two photons of equal energies going in opposite directions?

Comment: This case can be extended to real cases such as the Sun radiation, and shows that the distinction between mass and radiation is not relevant at the system level.

Problem 5.5: The Relativistic Momentum Solves the Problem of the Inelastic Collision in Section 4.1.2 in an Amazing Way

In the completely inelastic collision scenario of Section 4.1.2, we saw that the classical momentum definition is not compliant with the inertial frames equivalence postulate. Hence, we will redo the same calculations, but with the relativistic Momentum. We will see that the problem will be solved, but with an amazing effect: the masses of A and B must increase in the same proportion as their kinetic energies diminish. We will assume that these objects do not exchange thermal energy with their surroundings, knowing that this is only a theoretical hypothesis. (However, thermal processes are slow compared with a sudden collision.)

Q1: Express the spatial part of the relativistic system Momentum conservation law in K. Is it verified?

Q2A: Express the spatial part of the relativistic system Momentum in K′ after the collision. We will denote by m′ the objects masses after collision, but we will first assume that m′=m, meaning that the masses don't change during the collision.

Q2B: Express the spatial part of the relativistic system Momentum in K′ before the collision.

Q2C: What do you conclude? Calculate the mass change during the collision which allows the relativistic Momentum conservation in K′ to be respected.

Q3A: Express the system energy conservation in K. What is the amount of mass change that allows the relativistic energy conservation in K to be respected?

Q3B: Same question in K′.

Q4A: What is the kinetic energy variation in K?

Q4B: What is the kinetic energy variation in K′? Please comment.

Q5A: Express the system mass and show that it is constant in K.

Q5B: Same as Question 5A, but in K′, and show that it is invariant.

Problem 5.6: An Inelastic Collision

This is a continuation of Problem **4.5**: an object A of mass 2 kg is going at the speed 0,35 c along the OX axis of an inertial frame K. Another object B of mass 1 kg is going at the speed 0,7 c in the opposite direction. They will have an inelastic collision whereby both objects will be stuck together, forming an object C. We give: gamma (0.35 c)=1.067 and gamma (0.7 c)=1.400. Question: Will the mass of the object C be equal to the sum of the masses of A and B before collision? What will be the speed of C?

Problem 5.7: Example of Mass changes after an Inelastic Collision

The following example shows that in our day-to-day life, mass changes are negligible: two objects A and B of identical mass m=1,000 kg have an inelastic collision. Their speeds before collision are identical and equal 36,000 km/h, in opposite direction. What will be their masses m′ after the collision?

Problem 5.8: Atom Emitting a Photon

An atom of mass m is at rest in an inertial frame K, and spontaneously emits a photon of energy φ.

Q1: Does this emission make the atom lose some mass? Does it make it move? Does its energy change?

Q2: Show that the atom speed will be: $v \approx c \dfrac{\varphi^2}{mc^2}$ (1+ φ/mc^2); its

mass loss: $c^2 dm \approx \varphi + \dfrac{\varphi^2}{2mc^2}$. What is its energy variation? You

may use the classical kinetic energy approximation since the speed of the atom will be very low. Advice: use fully natural units, which will simplify the calculations.

Q3: Calculate the recoil speed of an atom of carbon which emits a gamma ray of 3.90 MeV, with its final mass being 12 u. Check that the classical kinetic energy approximation was justified. What part of the atom mass loss has been converted into the photon creation, and into the kinetic energy of the recoil?

Problem 5.9: Ice Melting and Impact on the Water Mass

Ice requires energy (from the atmosphere) in order to melt, the amount being 3.35×10^5 joules for 1 kg of ice to melt. What is the mass of the resulting water after 1,000 tons of ice has melt?

Problem 5.10: Objects' Speeds after an Elastic Collision

Consider an elastic collision between two objects A and B, their masses being, respectively, Ma and Mb. Before the collision, the speed of B relative to A is Sb. After the collision, the speed of B relative to A is S2b (Figure 5.8).

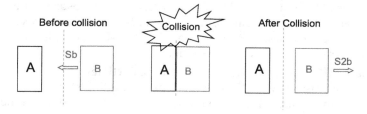

FIGURE 5.8
Elastic collision.

Q1: What speed must an object of mass m have so that its kinetic energy equals its intrinsic energy (i.e., its energy in its own frame)?

Q2: What speed must an object of mass m have so that its kinetic energy equals ten times its intrinsic energy?

Demonstrate that after the collision, the speed S2b is the opposite of Sb. To this end, use the Center of Momentum frame (denoted by CoM).

Problem 5.11: The Speed such that the Kinetic Energy equals the Energy due to the Mass

Q1: What speed must an object of mass m have so that its kinetic energy equals its intrinsic energy (i.e., its energy in its own frame)?

Q2: What speed must an object of mass m have so that its kinetic energy equals ten times its intrinsic energy?

Problem 5.12: Estimation of the Sun Lifetime

The Sun luminosity corresponds to a radiated power estimated to be: 3.85×10^{26} W. The total Sun mass is estimated to be 2×10^{30} kg. If we assume that 10% of its mass is hot enough to trigger thermonuclear reactions, how long does the Sun still have to radiate, assuming its luminosity remains constant?

Problem 5.13: The Carbon-12 Binding Energy

Carbon-12 is composed of 6 protons and 6 neutrons. Its mass is 12 u exactly. Is it stable? You may use the mass values given in Section 5.2.4. What is the minimum energy required to break the carbon-12 into its nucleons? (We neglect the effects due to the electrons.) What is the name for this energy?

Problem 5.14: The Decay of Beryllium

The beryllium nucleus has four protons and three neutrons. Its mass is 6,536 MeV. What is its binding energy? (Use the nucleons' masses of Section 5.2.4.) Then, a beryllium at rest absorbs a neutron, also at rest (approximately), which makes the beryllium decay into two alpha particles, whose rest mass is 3,728 MeV. What is the kinetic energy given to each alpha particle?

Problem 5.15: Energy of Ultra-Relativistic Particles

An "ultra-relativistic" particle is a particle going at a speed close to c; its gamma factor is important. Its $\beta = v/c$ is very close to 1; hence, we define: $\varepsilon = 1 - \beta$.

> **Q1:** Give a first-order approximation of the gamma factor, and then the energy relation.

> **Q2:** At the CERN particle accelerator, electrons reached the speed of 0.999 999 999 983 c. What is the energy of such an electron, knowing that its mass is ≈ 0.5 MeV/c^2? What is the percentage of the energy due to the mass part out of the total energy?

Problem 5.16: The acceleration due to the Newtonian-like Force

An object of mass m is at rest in an inertial frame K. Then from the time t of K onward, a Newtonian-like Force F is applied to this object. What acceleration does it take at the time t? Then the same Force is maintained on the object. What is its acceleration when its speed is 0.9 c? Use gamma (0.9 c) = 2.3.

Problem 5.17: The Fission Condition

Consider a nucleus A having its B/A ratio denoted by R. It is composed of two nuclei A_1 and A_2 whose B/A ratio, respectively, are R_1 and R_2. Show that if both R_1 and R_2 are greater than R, the nucleus A can incur fission.

Notes

1 Hence, the wording "equivalence energy-inertia" is preferable to "mass-energy", as further discussed in § 5.2.5.

2 The positron is a particle with the same mass as the electron, but with the positive charge e.

3 However, if the system is not isolated, part of the thermal energy generated by the collision is transferred to the objects' surroundings.

4 More information is given in Volume II §3.6.1.

5 Cf. more in Volume II §7.2.

6 Cf. §3.5.3.

7 Cf. Volume II §1.5.3.1.

8 This matter is called "black body" because it indiscriminately absorbs and emits all frequencies: it has no color.

6

Introduction to General Relativity

Introduction

First, the principle of equivalence will be carefully examined, as it is the cornerstone of General Relativity. Then two important consequences will be precisely explained: the bending of light in the neighborhood of great masses and the gravitational time dilatation effect. This will lead us to review and clarify the postulates of Relativity. Next, the way space-time deformations are mathematically characterized will be explained in a detailed and step-by-step approach. Then the principles leading to the famous equations of General Relativity will be explained. The first applications will finally be presented in a qualitative manner: the Mercury perihelion precession and the Schwarzschild's solution which enables us to characterize the bending of light, and thus provided a stunning confirmation of General Relativity.

6.1 The Principle of Equivalence

Special Relativity was a conceptual revolution which brilliantly explained many phenomena in a simple and unified way, but excepting an important domain: gravitation. Indeed, since Newton, gravitation was thought to produce instantaneous interactions, even over huge distances across the universe, whereas Special Relativity implied that nothing, even distant interactions, can go faster than light speed. For Einstein, this exception was not acceptable: the principles of Special Relativity were universal and not limited to electromagnetism, which was its originating domain.

Newton's gravitation theory actually left open several issues since its very beginning: for instance, if someone moves a big rock on Earth, how can the Moon instantaneously know that the center of mass of the Earth has changed? Can there be any instantaneous information transmission, and without any physical support? More generally, how can the Moon know where to fall? All these issues induced Newton to ironically say that one must be mad to believe in his gravitation theory.

Another mystery was the identity between the gravitational mass and the inertial mass. This issue was opened by Galileo who noticed that all objects fell at the same speed whatever their masses, but no explanation had been found for this coincidence.

Einstein issued in 1907 the principle of equivalence, "the happiest thought of my life[1]" (Einstein). It then took him 8 years of hard and complex work to translate this simple and luminous principle into equations, which not only revolutionized gravity, but again our vision of space, time, matter and energy. In 1919, his new theory was confirmed by the observation of the bending of light in the neighborhood of the Sun.

6.1.1 Statement of the Principle of Equivalence and first Consequences

> If you are inside a free-falling laboratory, all physical laws are those of Special Relativity.
>
> *Einstein*

The principle of equivalence means that a uniform gravitational field is equivalent to an acceleration. Everyone has experienced the effects of an elevator: for instance when it accelerates downward, you feel an inertial force lifting you upward. If someone cuts the cables holding the cabin, it suddenly becomes free-falling with the acceleration g (g=9.81 m/s² on Earth). Inside this free-falling cabin, you feel weightless and you actually are. Moreover, inside this cabin, you can observe that the laws of Special Relativity apply as if you were in an inertial frame and without gravity. For instance, if you give an impulse to a ball inside your cabin, it follows an inertial trajectory (constant velocity along a straight line). If you just leave a ball without giving it any impulse, it will remain fixed relative to you.

In classical physics and even with Special Relativity, this was explained by the presence of an inertial force, $\vec{f_i}$, caused by your acceleration and which offsets your weight since: $\vec{f_i} = -m\,\vec{a} = -m\,\vec{g}$.

With General Relativity, Einstein explained that a great mass deforms the chrono-geometry of our 4D space-time universe, and that these deformations produce accelerations which are perceived as gravity. Hence, there is no gravitational force, and consequently no inertial force that offsets it. The nonexistence of such an inertial force is due to the fact that the inertial frame actually is the free-falling cabin, and not the ground frame. The free-falling cabin indeed is inertial, being submitted to no force, whereas the ground incurs forces that prevent it from falling.

This principle has important immediate consequences:

- The mystery of the identity (or rather proportionality) between the inertial mass and the gravitational mass is solved: there is no gravitational mass since there is no gravity. Besides, Special Relativity showed that the quantity which opposes any acceleration relative to an inertial frame is not the classical mass, but the inertia (mγ).

- The issue about instantaneous distant interactions is also solved: Einstein indeed stated that great masses produce space-time deformations which propagate at the speed of light, which is finite.

The principle of equivalence has two other important and surprising consequences:

- Light must bend in the vicinity of very massive objects.
- Time must be influenced by very massive objects.

We will carefully explain these consequences, but before that, we will present a significant limitation of the principle of equivalence: if the gravitational field is not uniform, this principle is only valid locally.

6.1.2 Limitation of the Principle of Equivalence: The Tidal Effect

Looking precisely at the equivalence principle, we can notice that the gravity vector $\vec{g}$ is not exactly identical everywhere inside the free-falling laboratory because:

- $\vec{g}$ is directed toward the center of the important mass, but if you are on the left side of the laboratory, your gravity vector will not be parallel to that of another observer standing on the right side, whereas the laboratory acceleration is parallel to the same direction everywhere inside the aboratory as shown in Figure 6.1.
- $\vec{g}$ doesn't have exactly the same magnitude whether you are on the floor of the laboratory, or close to its ceiling since it is a function of the altitude, whereas the laboratory acceleration is again the same everywhere inside the laboratory.

Consequently, the equivalence gravity-acceleration is absolutely valid in one point of the laboratory only, but in all other parts, it is only a good approximation.

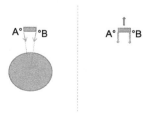

FIGURE 6.1
The difference between the effects of acceleration and gravity.

When the distances involved are such that the gravitational field significantly varies, its variations can produce important effects such as the tides of our oceans which will be further explained. Hence, these effects were called "tidal effects" even if they concern other aspects, like for instance in black holes where they are very important.

6.1.2.1 The Tidal Effect on Earth

Oceans are mainly attracted by the Earth, but also by the Sun and then by the Moon. Let's, however, focus on the Moon and we will see why: the distance Moon-ocean is not the same whether the ocean is on the side of the Earth facing the Moon, or on the opposite side of the Earth, the difference being the diameter of Earth. Consequently, oceans on the two opposite sides incur different gravitational attraction from the Moon, this difference being in the range of 6%. The same phenomenon applies to the Sun, but the relative gravitational difference is much less since the Sun is much more distant. Let's then return to the Moon attraction: the ocean which is closer to the Moon tends to have a more curved trajectory toward the Moon than the center of the Earth, so that there is a force lifting this ocean; conversely, the ocean which is in the opposite side of the Earth tends to have a less curved trajectory than the center of the Earth, so that it is also lifted (Figure 6.2).

When taking into account the Earth rotation relative to the Moon, we understand why there are two tides per day: if at 8 AM you are facing the Moon, the ocean is high; and then at 8 PM you will be in the furthest position relative to the Moon due to the Earth rotation: the ocean will again be high. When considering the similar tide effects due to the Sun, we understand why the tides are more important when the Sun and the Moon are aligned, and even more so when the Earth is closer to the Sun.

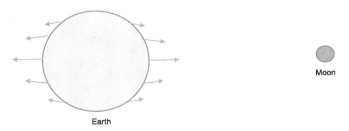

Moon

Earth

FIGURE 6.2
The tidal effect.

6.2 Important Consequences of the Principle of Equivalence

We will first see the stunning implications of the principle of equivalence on space: the bending of light near great masses; and then on time.

6.2.1 The Bending of Light Near Very Massive Objects

The following simple scenario will enable us to understand this phenomenon:

Imagine that you are on the last floor of a skyscraper, in front of a lift with a transparent cabin. You emit a light pulse horizontally toward the cabin: this pulse hits the cabin, goes inside, then exits outside the transparent cabin and finishes against the wall at a point D. Everyone will say that the point D is at the same height as your light pulse emitter, since the pulse was horizontally sent. However, Einstein showed that this is wrong, and we will see why:

We know from the equivalence principle that for a free-falling observer, everything is weightless, and it is like being in an inertial frame without gravity and where the laws of Special Relativity apply. Hence, let's do the following: at the very moment when the light beam touches the front side of the transparent cabin, a mechanism is triggered so that the cabin starts free-falling (e.g., someone cuts the cables).

An observer inside this free-falling cabin sees this light beam moving along a straight line since the laws of Special Relativity apply for him. The picture below illustrates this scenario (Figure 6.3). Let's denote by:

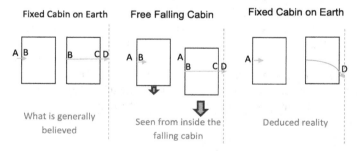

FIGURE 6.3
The bending of light.

- A: the pulse emission point.
- B: the point where the pulse first hits the cabin: we assume that B is at the middle of the cabin height.
- C: the point where the pulse leaves the cabin.
- D: the point where the pulse touches the wall after leaving the cabin.

The cabin had no speed when the beam touched its front side on the point A, so that the observer inside the cabin sees this beam moving horizontally, meaning for him in a perpendicular direction relative to the cabin wall. The pulse will then hit the right side of the cabin at the middle of its height: this means that the point C is at the middle of the cabin height, like B.

The cabin being transparent, the beam continues outside the cabin, and reaches the point D on the skyscraper wall. D is below A since the lift has been moving downward while the pulse has been traveling inside the cabin.

The cabin being transparent, the light pulse actually has the same trajectory as if there were no lift. (The motion of the air inside the cabin has no impact on light propagation, nor the motion of the glass walls.) Consequently, the position of the point D is on the light pulse trajectory even if there is no cabin, and it is clear that D is below A.

We used the fact that our free-falling observer was legitimate to state that light propagates along a straight line, thanks to the equivalence principle, whereas fixed observers on Earth did not have this right, since they are not free-falling. (They should have said: we don't really know the influence of gravity on light propagation.)

Thus, gravity bends light; we can also say light falls like any object, but in such a minute manner that it was undetectable. For example, a horizontal light beam 1 km long incurs a deflection in the range of 10^{-18} m. Hence, we needed much more massive objects than the Earth to detect such phenomena, and it was in 1919, taking advantage of a solar eclipse that observations confirmed the bending of light coming from stars in the neighborhood of the Sun. The observed deflection angle was very close to the 1.75 arc-seconds predicted by General Relativity, which brilliantly confirmed this new theory.

Remark 6.2: If it were a bullet instead of the light pulse, the same reasoning would show that the point D would be lower since the bullet speed is lower than c.

Remark 6.3: Like in classical physics, light follows the path which extremizes the distance covered. The fact the light bends in the neighborhood of a great mass means that our space-time universe is not Euclidean, but incurs chrono-geometrical distortions generated by the great mass in its vicinity (cf. more in Section 6.3).

Having seen the impacts of the principle of equivalence on space, we will now see its impacts on time.

6.2.2 Impacts of Gravity on Time

6.2.2.1 The Proper Time Universality

The proper time universality plays an important role: the clock postulate states that the proper time is not affected by the clock's acceleration (cf. Section 3.5.2). Now the principle of equivalence implies that the proper time of a clock must also not be affected by gravity since there is no gravity, but only accelerations.

One consequence is that the official time definition of the second, which is based on a specified amount of radiation of an atom of cesium-133, applies everywhere, even on very massive planets and with the same local perception. This important feature was confirmed by experiments.

Besides, we saw that the very metric of our space-time continuum is based on the Lorentzian norm invariance, which reinforces the importance of the clock postulate (we recall that $ds^2 (dM) = c^2 d\tau^2$.).

However, the universality of the proper time doesn't mean that time is not affected by gravity, as we will see now:

6.2.2.2 The Gravitational Time Dilatation

We will first wonder if two observers located at different altitudes see the clock of the other observer beating at the same pace as his own clock. The following scenario will give us the clue:

Consider two observers A and B on a same vertical line on Earth, with A being higher than B. We will assume that the length H of the segment AB is small enough that gravity is constant. We will suppress all gravitational effects by again imagining a transparent free-falling cabin (Figure 6.4):

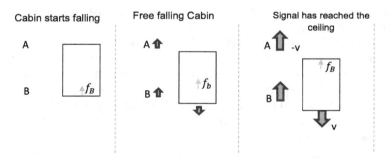

FIGURE 6.4
Impact of gravity on the time.

Initially, the cabin is fixed relative to the earth with its floor being at the level of B and its ceiling at the level of A. Then at the precise moment when the cabin starts free-falling, a signal with the frequency f_B is emitted from the cabin floor upward toward its ceiling.

Seen by B, this signal is seen with the same frequency f_B because the cabin speed is null when the signal is emitted. Then during the free-falling motion of the cabin, an observer R inside this cabin can say that the signal is moving upward always with the same frequency f_B, since he can apply the laws of Special Relativity, being a free-falling observer. (R is in a referee position.)

Then, at the time when the signal reaches the ceiling, the cabin has acquired a certain velocity v, so that the observer R can say that the observer

A is moving upward with the velocity −v, and consequently, he sees the signal with a frequency f_A which is lower than f_B due to the Doppler effect (cf. Section 2.3.2). The frequency seen by A is indeed lower because A is moving away from the signal.

Calculations: The time duration T measured in the cabin frame K for the pulse to reach A is T=H/c. During T, the velocity acquired by the cabin is $v=gT=gH/c$. This velocity is the opposite of the velocity of A seen from K, and which must be taken into account for the Doppler effect incurred by A; hence, we have: $f_A = f_B \cdot \sqrt[2]{1-v/c} / \sqrt[2]{1+v/c}$. In usual cases, v ≪ c, so:

$$f_A \approx f_B\left(1-\frac{gH}{2c^2}\right)\left(1-\frac{gH}{2c^2}\right) \approx f_B\left(1-\frac{gH}{c^2}\right) \blacksquare \qquad (6.1)$$

Symmetrically, a signal emitted by A toward B will have its frequency increased when received by B due to the same Doppler effect, but this time, R sees B coming toward the signal.

6.2.2.2.1 Implication Regarding the Time: The Gravitational Time Dilatation Effect

There is a relationship between the frequency f_B and the proper time of B. The signal has f_B periods during 1 second of the proper time of B, so that each period lasts $1/f_B$ second of the proper time in B.

Similarly, one period of the signal received in A lasts $1/f_A$ second of the proper time in A, and we just saw that: $f_A < f_B$.

For the sake of simplicity and clarity, let's assume that $f_B = 100$, $f_A = 50$ and that B emits his signal during 1 second of his proper time, and then stops emitting. Thus, one period of the signal emitted lasts 0.01 second of the proper time in B, and B will emit 100 periods. The signal will then pass in front of A, and the time taken by these 100 periods to pass in front of A will give the perception that A has of 1 second of the proper time of B. A actually sees 50 periods per second; hence, the 100 periods emitted by B will need 2 seconds to pass in front of A. A thus sees 1 second in B lasting 2 seconds on his watch. Thus more generally, **A sees B's proper time dilated** by the factor: $\sqrt[2]{1+v/c} / \sqrt[2]{1-v/c}$.

When v ≪ c, this factor can be approximated by: $1+\frac{gH}{c^2}$.

Symmetrically, B sees the time of A running faster. There is no contradiction between the perceptions of A and B, each of them can agree that **anyone sees an upper clock running faster than a his own clock** (and slower for a lower clock).

This gravitational time dilatation effect not only is a perception effect, but has real impacts: for instance, if a space traveler undergoes a journey in which he stays several years on a very massive planet, he will be younger than his colleagues when he returns to Earth. Indeed, let's assume his stay on the massive planet B is long enough, for instance 20 years, so that we can neglect the effects of the time for going to this planet and returning to Earth. The previous scenario shows that his proper time is seen dilated on Earth: each of his seconds lasts more than 1 second on Earth. After 20 years on B, the time elapsed on Earth is much longer than 20 years. If he had a twin on Earth, he will be younger than him at his return. Note that the traveling twin, during all his trip, hasn't noticed any change regarding the way time flows: his heart has always been beating at the same pace (except for the possible vibrations of the rocket), and he continues to incur aging effects at the same pace.

This gravitational dilatation effect adds to the effect on the time due to the accelerations, which were seen with Langevin's twin paradox: this traveler now has two reasons to be younger than his twin who remained on Earth.

Remark 6.4: A common mistake is to conclude that time passes more slowly at a greater gravity level than at a lower one: indeed, the proper time is universal. Someone in the position of the observer A should say that he **sees** the time in B passing more slowly than his own time. This situation is similar to Langevin's Travelers paradox: it would be wrong to say that time passes more slowly for the accelerated traveler because he is younger than his inertial twin when they meet again. Again, their proper times always beat at the same pace. These so-called dilatation effects are actually due to time de-synchronization.

Remark 6.5: This gravitational time dilatation law plays an important role in the GPS (Global Positioning System): it yielded an improvement in the system accuracy by a factor of 10^4! Otherwise, the GPS precision would have been of few kilometers, instead of the few meters as today. The GPS indeed involves several satellites, and the proper time of a clock inside a satellite is seen beating faster on Earth.

Remark 6.6: In cosmology, this gravitational time dilatation effect adds to the redshift due to the universe expansion, and this has been confirmed by observations: light emitted by very massive objects shifts more to the red. However, this effect is usually small: for instance, an observer located very far from the Sun would see a perfect clock on the sun surface beating slower than his own by 2.10^{-6} seconds, which represents 64 seconds in 1 year. A clock in the center of the Sun would be seen losing 5 minutes per year. Conversely, a perfect clock on the horizon of a black hole would be seen beating infinitely slowly.

Remark 6.7: As an order of magnitude, an altitude difference of 1 meter gives a time difference of 10^{-16} seconds, which can be detected, thanks to the very high precision of our best atomic clocks.

6.2.2.2.2 Consistency with Photon Energy Variation

The previous scenario invites us to further examine the case of the photon emitted from B with the frequency $f°$ upward toward A. We indeed found that this photon is seen in A having the frequency f, with $f < f°$. This frequency variation also means energy variation due to the relation $E = hf$ (h being the Planck constant), and this energy variation should be explained:

The photon has both a wave and a corpuscle nature. According to its wave nature, the photon energy variation between B and A is equal to: $dE = h(f - f°)$. According to its corpuscle nature, and to the energy-inertia equivalence, the photon has an inertia, denoted by m*, which is: $m* = E/c^2 = hf°/c^2$. The photon emitted from B needs some energy to go upward against the gravitational field, and the work of the gravitational force is equal to its energy variation[2]. We thus have: $m^*gH=h.(f_B-f_A)$; then: $h.\,f_B.g.H/c^2=h.(f_B-f_A)$, so:

$$\frac{(f_B - f_A)}{f_B} = \frac{gH}{c^2}$$

(6.2)

which is equivalent to equation 6.1.

We can say that when going upward and struggling against a gravitational field, a photon loses energy in the form of a reduction of its frequency. Symmetrically, when falling down in a gravitational field, a photon acquires more energy, and so its frequency increases.

In Volume II Section 3.6.6.3, the relation 6.2 is integrated; moreover, the gravitational potential, denoted by Φ, is used, which yields:

$$\frac{f_B}{f_A} = \exp(\frac{\Phi_A - \Phi_B}{c^2})$$

(6.3)

which for small potential variations can be approximated by:

$$1 + \frac{d\Phi}{c^2}.$$

(6.4)

The relations 6.3 and 6.4 are valid using the gravitational potential of General Relativity, but for low gravity such as on the Earth, the Newtonian potential, i.e., $\frac{GM}{R}$, is a very good approximation.

Remark 6.8: The reader may be confused since the principle of equivalence states that there is no gravitational force. This point will be clear with the next chapter: in summary, the observer in A is not in an inertial frame; consequently, he sees every object having an inertial force, similarly as in classical physics.

6.2.3 Revisit of the Fundamental Postulates and Concepts

The concept of inertial frame on which the whole Special Relativity is grounded has been undermined by General Relativity: it relies on the existence of straight lines, and the latter rely on the law of inertia stating that an object submitted to no force follows a linear trajectory with constant velocity. However, we saw that this law is not true since light bends in the neighborhood of a great mass.

However, according to the principle of equivalence, in the vicinity of any event, the laws of Special Relativity apply for a free-falling (inertial) observer. He sees photons going along straight lines in his vicinity. It is then possible to have a 4D Euclidean space in the vicinity of an event; this space is called the Euclidean tangent space (tangent to the curved space or curved surface).

In this tangent space, we can have an orthonormal frame of Minkowski, but it is not an obligation to choose orthonormal frames.

Regarding the new law of inertia, it can be found by reformulating the core of the previous law in a way which is intrinsic: **the proper speed of an inertial (free-falling) object is constant**. The proper speed can indeed be defined at each event M of an object trajectory, by the 4-Vector $\dfrac{dM}{d\tau}$ which belongs to the tangent space at that event.

In a curved surface, an inertial (free-falling) object follows a trajectory that respects the constancy of its proper speed. Mathematically, this is more complex to apprehend than in Special Relativity because in a curved surface, one cannot compare vectors that are not in the same Euclidean tangent space, and indeed two events have two different Euclidean tangent spaces. However, there exists a method which makes it possible (cf. 6.3.4 and Volume II Section 4.2), and shows that the inertial object follows a geodesic. The geodesic indeed has two characteristic properties: it extremizes the distance between two events, and it is always parallel to the same direction. However, the distance is now Lorentzian, and the notion of parallelism has to be adapted to curved surface. If the space is Euclidean, the geodesic is a straight line.

> There are some familiar examples of non-Euclidean spaces: for instance, the 2D space represented by the surface of a sphere like the Earth. Imagine that you are walking always along the same direction: you believe that you are walking along a straight line, but you will be surprised that after a certain time, you will be back at the same place where you started. A Euclidean straight line would not have permitted that. Another strange phenomenon will convince you that you are not in a Euclidean space: Let's have two points A and B on the equator; at each point, you draw a perpendicular line relative to the equator. These two perpendicular lines will meet at each pole, which contradicts another Euclidean postulate (parallel lines never meet).

As laws of Special Relativity apply in the tangent space, we still have the Lorentzian metric (distance) in this space, with the property that the distance between two events M and N forming a time-like interval represents the proper time taken by an inertial object to go from M to N: $c.\tau_{MN} = \sqrt{ds^2(MN)}$. The trajectory of an inertial object being a geodesic, it means that this object chooses the trajectory that locally maximizes its trip duration, as measured with its proper time.

The clock postulate, meaning the proper time universality, takes an even greater importance than in Special Relativity. Let's illustrate some effects of this postulate: we build several identical perfect clocks in one place, and then send them in different locations with different speeds, accelerations and even gravitational fields. These clocks will continue to beat at the same pace as seen by observers co-located with them; however, they won't be synchronized.

According to the principle of equivalence, the speed of light speed is still c, but locally. The relation d=ct shows that the proper distance is also universal, but locally only. We can then have units of time and distance which are phys-ically the same everywhere (e.g., the second given by a perfect clock and the meter given by the wavelength of given radiation). Of course, these units are seen differently by moving observers or in regions with different gravity levels.

The word "local" (or "locally") is frequently used in General Relativity; hence, it is important to make precise its meaning: "local" means that the limit of the difference between the actual results of a physical phenomenon taking place in the neighborhood of an event E and its prediction according to the laws of Special Relativity applied in the Euclidean tangent frame in E, tends to zero when the Lorentzian distance of this phenomenon to E tends to zero. In reality, locally can mean a very large region if we are very far from a very massive object, or extremely small if we are very close to extremely massive ones like for black holes (cf. Section 7.2).

In addition, another conceptual problem with perfect inertial frames is that their physical reality is based on rigid objects. However, absolutely rigid objects cannot exist since they would contradict the existence of a speed limit: if you push one end of a "rigid" ruler, and if the other end reacts imme-diately, it would represent an immediate transmission of information, which is impossible. What actually happens when you push one end of the ruler is that you generate a shock wave within the structure of the ruler which propagates at a speed far below c.

Still, the word "inertial frame" is commonly used in General Relativity, but with an important difference with Special Relativity: the core meaning of the word "inert" remains valid, which is "incurring no force", but in General Relativity, an inertial frame is a free-falling frame, and no longer a frame evolving along a straight line with constant velocity (except when it is very far from great masses).

Like in Special Relativity, all inertial (free-falling) frames constitute a spe-cial class of frames: an observer can tell if his frame is inertial by the follow-ing simple experiment: he holds a ball, and then lets it free without exerting any impulse. If he sees the ball staying fixed relative to him, then his frame is inertial; otherwise, it isn't.

> *The principle of Mach:* E. Mach, who was a strong opponent to the abso-lute frame, saw in the class of inertial frames a privileged class which reminds of the privileged status of the Newtonian absolute frame. He then deduced that the existence of this privileged class of frames frame is due to the presence of distant great masses in the universe, and that the inertial forces were due to these distant masses.
>
> This principle induced Einstein to state that space and matter are linked together: masses have created the space, and a space without mass cannot exist. The wording space-time deformations or chrono-geometri-cal distortions may then be confusing as one may think that there pre-existed a 4D Euclidean frame before the existence of the masses, and that the masses have distorted it. This vision also has the same drawback of

the Newtonian absolute frame. Hence, our 4D universe should be considered as it is: a dynamic entity with its masses. Even when addressing the space expansion, one should not consider that space expands independently of the masses. However, for the sake of simplicity, we will still use this wording "deformation", while being aware of its limitation.

Non-inertial frames are called "accelerated" frames. Like in Special Relativity, an object in an accelerated frame incurs an inertial force, but in General Relativity, the gravitational force is an inertial force. The following example illustrates this point:

Imagine you are on a cliff and you see an object falling: this object is inertial since it is subjected to no force. Conversely, you are sustained by the cliff which exerts a force that keeps you still, which means that you are not inert and so your frame is an accelerated frame (whereas you thought it was fixed). Then, being in an accelerated frame, with an acceleration of $g \approx 9.806$ m/s^2 (function of your altitude), you will consider that all objects incur inertial forces, and you will explain the acceleration taken by the free-falling object relative to your frame by the inertial force it incurs. Hence, the force of gravity is said to be fictitious (or apparent), and the same for inertial forces.

Besides, the fundamental system Momentum conservation remains valid in General Relativity (for isolated systems), in accordance with the principle of equivalence and the universal homogeneity, with in particular the invariance of physical laws with time.

The fundamental axiomatic pillar thus appears to be the universal homogeneity and isotropy. Having seen that we can have units of time and distance which are locally the same everywhere, the universal homogeneity and isotropy imply that any local physical experiment gives the same result in any inertial (free-falling) frame. This is an extension of the principle of Relativity, hence the name General Relativity.

Moreover, after the theory of Relativity was established, it has been shown that the light speed invariance postulate is not necessary: it is indeed a consequence of the universal homogeneity in conjunction with the reject of the absolute frame (cf. Volume II Section 7.3.1). Hence, we are induced to further examine the homogeneity postulate:

Homogeneity regarding extremely large scales: Over very long distances, greater than 10^{19} km, it appears that the quantity of galaxies is the same in all directions and at all distances. Over short distances, the universe is obviously not homogeneous. Moreover, our universe incurs an expansion without privileging any direction, and this expansion is currently accelerating also without privileging any direction. The geometry of the universe seems to have no curvature: it is said to be "flat". Moreover, the Cosmic Microwave Background, which gives a picture of the universe soon after the Big Bang, shows a remarkable homogeneity and isotropy. Thus in the extremely large scale, our universe appears to be homogeneous, isotropic and flat. It is not simple to represent such space; it is not like a sphere since a sphere has a privileged point, its center. Conversely, in our flat universe, any point can be considered as its center. However, the homogeneity in time is an open issue: the universe has incurred several phases of expansion, which are not fully explained. Moreover, we know very little about the mysterious presence of black and grey matter which constitute no less than 95% of the overall matter!

Regarding extremely small scales: There is a minimum scale for distance and time below which our physical laws don't apply. These are the Planck distance and the Planck time; they are linked by the relation $d = ct$. One implication is that we cannot say anything about what happened during the first 10^{-44} seconds after the Big Bang. The event and the point object scheme thus have theoretical limitations, even if we are very far from reaching them: as of May 2010, the smallest uncertainty time interval in direct measurements was in the order of 1.2×10^{-17} seconds, which is about 2.2×10^{26} Planck times.

Besides, the event scheme on which Special and General Relativity are built has some limitations: the photon, for instance, doesn't have a proper time, and one cannot tell where it is located, except in some cases. Moreover, L. de Broglie showed that not only the photon, but all particles behave like waves under certain conditions, which was confirmed by experiments where electrons could interfere with themselves (C. Jönsson). Furthermore, the Heisenberg principle states that the more precisely the position of a particle is determined, the less precisely its momentum can be known (and vice versa). The reason for this impossibility to precisely know the trajectory of an elementary particle is that it doesn't have any trajectory: its presence is depicted by a cloud of probabilities in accordance with Schrödinger's equation. At extremely small scales, quantum mechanics and Relativity show important incompatibility, and this still is an important research domain of physics.

6.3 Curved Surfaces: Distance and Curvature Characterization

We saw that the presence of masses/energy distorts our space-time universe, producing effects which were called gravity. We will first see how distortions can be mathematically characterized, and then how these distortions can be related to their causes, which will finally result in the famous equations of General Relativity.

The mathematical subject of curved space has been studied first by Gauss, then Riemann, Christoffel, Levi-Civita and Ricci. Einstein adapted these concepts to our deformable space-time universe with the help of his friend M. Grossmann, and succeeded in November 1915 to find the equations of General Relativity.

We will follow a similar outline as T. Damour in his excellent book "Once upon Einstein". We will see how the curvature of a surface can be characterized, and how the notion of distance can be defined on such surface. We will first examine these questions in the simpler context of real 2D surfaces, and then of 3D volumes, and finally, we will adapt these concepts and results to our 4D space-time universe with its Lorentzian metric.

6.3.1 Distance in 2D Surfaces

We will first recall the distance definition on a plane surface, and see its main properties. We will then imagine a plane surface having some elasticity so that we will distort it and see the impacts on the distance between two points. This will also enable us to characterize the surface curvature.

6.3.1.1 Distance in Simple 2D Plane Surfaces

Consider a surface which initially is plane. We can draw on this surface a set of regularly spaced parallel lines in one direction, and another set of regularly spaced parallel lines in a perpendicular direction. These two sets of lines draw a regular network of small squares, forming an orthonormal system of coordinates, denoted by K, as shown in Figure 6.5 below which presents the plane surface from a perspective. (Hence, the right angles are not seen right, but they are.)

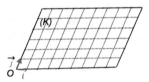

FIGURE 6.5
Plane surface with an orthonormal grid.

In classical geometry, distance is an intrinsic notion which can be calculated, thanks to Pythagoras: the distance^2 between any couple of points (A, B) on the surface is:

$$AB^2 = (X_B - X_A)^2 + (Y_B - Y_A)^2. \tag{6.5}$$

Had we chosen another orthonormal frame with different axis directions, the result of this calculation would be the same, provided that the lengths of the unit basis vectors were identical in both frames.

However, this Pythagorean relation is valid only if the frame is orthonormal, but as we will address curved surfaces, we will encounter non-orthonormal frames. The norm^2 (or length^2) of the segment AB is the same even if the frame is not orthonormal, but it is not obtained by the equation 6.5, but by another way that we will see now:

Mathematically, any norm is a scalar product of a vector with itself:

$$AB^2 = \overrightarrow{AB} \cdot \overrightarrow{AB}. \tag{6.6}$$

This is also true in non-orthonormal frames. Consider any non-orthonormal base, with its unit vectors denoted by $\vec{i}$ and $\vec{j}$. Any vector $\vec{u}$ can be expressed with this base:

$\vec{u}=u_x\vec{i}+u_y\vec{j}$. The same for a vector $\vec{v}$: $\vec{v}=v_x\vec{i}+v_y\vec{j}$.

The scalar product $\vec{u}\cdot\vec{v}$ is:

$\vec{u}\cdot\vec{v}=(u_x\vec{i}+u_y\vec{j})\cdot(v_x\vec{i}+v_y\vec{j})=u_xv_x i^2+\left(u_xv_y+u_yv_x\right)\left(\vec{i}\cdot\vec{j}\right)+u_yv_y j^2$.

The norm² of the vector $\vec{u}$ is then:

$$\vec{u}^2=u^2=\vec{u}\cdot\vec{u}=u_x^2 i^2+2\left(u_xu_y\right)\left(\vec{i}\cdot\vec{j}\right)+u_y^2 j^2 . \tag{6.7}$$

Mathematically, this relation is called a "quadratic form", meaning that it is a polynomial of the second degree of the coordinates u_x and u_y. In the particular case where the basis is orthonormal, we have $i^2=j^2=1$ and $\vec{i}\cdot\vec{j}=0$, so that the equation 6.7 becomes identical to equation 6.5.

Also, we have: $\vec{i}\cdot\vec{j}=ij\cos(\vec{i},\vec{j})$, which shows that the parameters of the quadratic form reflect the geometry of the unit basis vectors.

6.3.1.2 Distance Calculation on a Curved Surface

6.3.1.2.1 The Issues When the Surface Is Curved

We assume that our initial surface has some elasticity and we distort it, for instance by placing a heavy ball on it as shown in the figure below. The lines on the surface are also distorted; they don't form squares any more, especially in the zones where the distortions are important: the squares become trapezoids and, consequently, our network of lines don't form an orthonormal frame any longer; we then have a problem for calculating distances.

The good news is that there exists a method to calculate the distance between any couple of close points (M, N) on the curved surface, and that this method will also enable us to characterize the surface curvature (Figure 6.6):

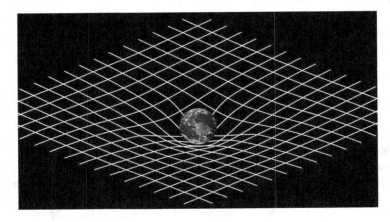

FIGURE 6.6
Curved surface – distorted grid. (Image by Lê Nguyên Hoang, published on Science4All.org.)

First, we notice that the network of curved lines still constitutes a system of coordinates: any point can be localized as being at the intersection of the Nth line with the Mth column (or close to it). The precision of this method can be as good as desired, by reducing the size of the squares of our initial network. We can take, for instance, 1 mm between for the initial square size, or even 1 micron if such precision is needed. This is called a curvilinear system of coordinates, which will be denoted by K'. Further, we assume that the surface is continuous and even differentiable.

6.3.1.3 Methodology for Distance Calculation on the Curved Surface

Consider a curved line on the curved surface between two points, A and B. This line successively crosses several trapezoids. Having drawn an extremely tight network of lines on our initial surface, the trapezoids on the surface are also very small. Call M the point where the line enters one very small trapezoid, and N the point where the line exits this trapezoid. Having assumed that the surface is differentiable, the portion of the surface within each trapezoid can be approximated with a plane, called the "tangent plane", which will enable us to calculate the length of the section MN within our small trapezoid.

The length of the curved line between A and B will then be obtained by summing up all the very small lengths in the succession of trapezoids that are crossed by the line AB.

It still remains to calculate the distance MN within one very small trapezoid:

Let's consider the plane K' which is tangent to the surface at the point M. The line MN can then be approximated by the straight line MN on the small trapezoid which is around M on K'. This will enable us to express the distance MN as follows (Figure 6.7):

(K')

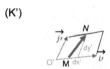

FIGURE 6.7
Distance in a small trapezoid.

In this tangent plane K', we choose the two unit basis vectors, $\vec{i}'$ and $\vec{j}'$, which are made of the two sides of the trapezoid, and we choose the origin O' as the intersection of these two sides. This frame is not orthonormal; still, the vector $\overline{MN}$ can be expressed with this basis: $\overline{MN} = dx'\vec{i}' + dy'\vec{j}'$. Subsequently, the distance MN can be calculated with equation 6.7:

$$MN^2 = \overline{MN} \cdot \overline{MN} = dx'^2 i'^2 + 2dx'dy'(\vec{i}' \cdot \vec{j}') + dy'^2 j'^2 \tag{6.8}$$

It is still a quadratic form, with its coefficients being i'^2, $2(\vec{i}' \cdot \vec{j}')$ and j'^2. These coefficients reflect the surface curvature: let's remember that before the surface was distorted, it was planar having a network of perpendicular lines forming an orthonormal frame. Then, the more the surface has been distorted, the more different are the trapezoids from their original squares, meaning the more different are their angles from 90°, and the more different are the lengths of the trapezoids sides from their original squares.

This relationship between these coefficients and the surface curvature can further be characterized as follows: let's consider the original orthonormal frame K, with its unit basis vectors being $\vec{i}$ and $\vec{j}$. We can add a third unit vector $\vec{k}$, perpendicular to this plane, with the same length as the two others (for instance, 1 mm) so that K is a 3D orthonormal frame. K will be our new reference frame, and we can express the unit vectors of K' as a function of the unit vectors of K: $\vec{i}' = a\vec{i} + b\vec{j} + c\vec{k}$ and $\vec{j}' = d\vec{i} + e\vec{j} + f\vec{k}$.

This enables us to calculate the scalar product: $\vec{i}' \cdot \vec{j}' = ad + be + cf$ (taking into account the fact that the frame K is orthonormal). The relation 6.8 becomes:

$$MN^2 = \overrightarrow{MN} \cdot \overrightarrow{MN} = i'^2 . dx'^2 + 2(ad + be + cf). dx'dy' + j'^2 . dy'^2 . \qquad (6.9)$$

Let's find the meaning of the parameters a, b, c, d, e and f: for this, we make the scalar product: $\vec{i} \cdot \vec{i}' = \vec{i} \cdot (a\vec{i} + b\vec{j} + c\vec{k}) = a . i^2 = a$. We took into account that K is orthonormal. Likewise, we have: $\vec{i} \cdot \vec{i}' = i \cdot i' \cos(\vec{i}, \vec{i}') = i' . \cos(\vec{i}, \vec{i}')$; hence: $a = i' . \cos(\vec{i}, \vec{i}')$.

The same reasoning applies to other parameters, and we find: $\vec{j} \cdot \vec{i}' = \vec{i} \cdot (a\vec{i} + b\vec{j} + c\vec{k}) = b$ with $b = i' . \cos(\vec{j}, \vec{i}')$, and so on for the other parameters.

The lengths i' and j' together with the angles $(\vec{i}, \vec{i}')$, $(\vec{j}, \vec{i}')$ and their variations in the neighborhood of the point M characterize the surface curvature. The knowledge of all these parameters at all points of the surface enables us to reconstruct the curved surface.

Remark 6.9: We have chosen the frame K to be the initial plane surface; we could have chosen any other orthonormal frame for our reference frame, provided that the new unit basis vectors have the same length. This is indeed due to the invariance of the norm (i^2) and (j^2), and also of the scalar product $(\vec{i}', \vec{j}')$; and this also shows that the curvature is intrinsic. On the other hand, had we chosen another curvilinear frame on the same curved surface, the parameters would have been different since the angle $(\vec{i}, \vec{i}')$ and the lengths of the sides of the trapezoid would have been different.

6.3.1.4 Distance Calculation Between Distant Points, and the Geodesic

Next, consider a line between A and B on the curved surface. We can calculate the length (distance) of this line by adding the lengths of the succession of all small segments in all small trapezoids that are crossed by this line.

We assumed the surface is differentiable, so that when the size of each trapezoid tends to zero, the above sum tends to a limit which is the length of the line between A and B. Moreover, in case the line AB can be described by a mathematical function, the length of this line can be obtained by a curvilinear integration along this line: $L(AB) = \int_{AB} dM$.

We mentioned that in General Relativity, an inertial (free-falling) object has its Momentum constant, $P = m.dM/d\tau = cst$. This implies that its trajectory is a geodesic in accordance with the characteristic property of the geodesic, which is a line where the tangent to any of its points is parallel to the same direction.

The second characteristic property of the geodesic is to be the line which extremizes the distance between any couple of its points.

If space is Euclidean, these characteristics define the straight line. In non-Euclidean spaces, these properties are generally complex to apply, and this will be further seen in the next chapter.

> A first example can be given in the case of a sphere: the geodesic between two points M and N on the surface of the sphere is the portion of great circle joining these two points. This line is indeed the shortest one between these points.

When the surface is more complex than a sphere, an inertial object going from A to B chooses a succession of local extrema, which may not result in the shortest path between two distant points.

6.3.2 Toward 4D: Case of 3D Deformable Volumes

The previous method can also be applied to deformable volumes: let's imagine a cubic block made of a substance having some elasticity, such as a jelly. We design inside this volume a network of perpendicular lines, regularly spaced, which represents a 3D orthonormal network. We can again enhance the precision by having a very tight network, such as 1 mm (or even 1 micron) for the side size of each small cube.

We then submit this block of jelly to a strong wind on the upper part of the block. Being flexible, the jelly will be distorted and each small cube will become a parallelepiped. We will show how to calculate the distance inside each small parallelepiped, which will enable us to calculate the distance of any line between any couple of points A and B inside this jelly: by adding up (integrating) the distances of the small segments in each parallelepiped that was crossed by the line AB.

Let's imagine the line AB crosses a parallelepiped, entering in M and exiting in N. We want to calculate the distance MN (Figure 6.8):

(K')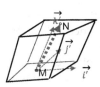

FIGURE 6.8
Distance in a small parallelepiped.

We can approximate the length of MN by assuming that it is a very small section of straight line. In the small parallelepiped, space is assumed to be Euclidean; it is the Euclidean tangent space to the point M. The length of the small segment MN can be calculated by applying equation 7.2 and extending equation 7.3 to 3D:

$$MN^2 = \overrightarrow{MN} \cdot \overrightarrow{MN} = (dx'\vec{i}' + dy'\vec{j}' + dz'\vec{k}') \cdot (dx'\vec{i}' + dy'\vec{j}' + dz'\vec{k}'), \quad (6.10)$$

where $\vec{i}', \vec{j}', \vec{k}'$ are unit basis vectors of the 3D Euclidean tangent frame in M; they have been chosen such that their directions and their lengths are identical to the ones of the sides of the small parallelepiped which are the closest to M.

When expanding equation 6.10, we obtain a quadratic form whose coefficients are classically denoted by $g_{\mu\nu}$:

$$MN^2 = \sum_{\mu\nu} g_{\mu\nu} dx'^{\mu} dx'^{\nu}. \quad (6.11)$$

Let's explain this notation as it is widely used in General Relativity: the terms dx', dy', dz' are denoted by: dx'^1, dx'^2, dx'^3. The letters μ and ν are indices taking successively the values 1, 2, 3. The maximum number 3 corresponds to the number of dimensions of the space. Similarly, the unit basis vectors $\vec{i}'$, $\vec{j}'$, $\vec{k}'$ are denoted by $\vec{i}_1, \vec{i}_2, \vec{i}_3$.

Each $g_{\mu\nu}$ coefficient is the scalar product between the unit basis vectors $\vec{i}_\mu$ and $\vec{i}_\nu$. The letter g was chosen because these coefficients reflect the geometry of the distorted block of jelly as seen previously (knowing that the same notions of angle and length apply in 3D).

There are $3 \times 3 = 9$ different $g_{\mu\nu}$ coefficients, which can be represented with a matrix of dimension 3×3. The scalar product being symmetric (commutative), this matrix is symmetric; hence, there are only 6 different coefficients.

6.3.2.1 Distorted 4D Space-Time Universe

Our space-time universe can be modeled by a block of jelly having some elasticity, and which is distorted by the presence of mass-energy. However, we now have a fourth dimension for the time, which also incurs distortions. Besides, the following adaptations are necessary: in the classical 3D space, the elementary unit was the point; now it is the 4D event; similarly, vectors need to be replaced with 4D space-time separation vectors. Another important difference with the 3D world concerns the distance definition, which is Lorentzian.

Remark 6.10: This model with a block of jelly is useful from a mathematical perspective, but physically, gravity does not need any substance to propagate, similarly as electromagnetism. Besides, this model conveys the idea that a Euclidean space pre-existed and that the presence of masses has distorted it, which is not consistent with the vision that matter-energy creates the space.

6.3.2.2 The Lorentzian Distance

The classical distance definition cannot apply because i) it ignores the time dimension, and ii) it is no longer intrinsic (distances being relative). Hence, the distance in our space-time universe is the Lorentzian one, and it is the only one.

Mathematically, any norm^2 is a scalar product of a vector with itself. Regarding the Lorentzian norm, the corresponding scalar product is the Lorentzian scalar product.

> *Notation:* As there is a risk of confusion between the classical scalar product and the Lorentzian one, we denote the latter by the symbol •. Besides, the 4-vectors are generally denoted without the arrow on the top, but in bold. We thus have: $ds^2(V) = V^2 = V•V$. Then when the reader is familiar with this notion and has no risk of confusion, he should use the commonly used notation, which is the same dot as the classical scalar product: $V·V'$ or even $V.V'$.

In a orthonormal 4D frame, such as the frame of Minkowski, the Lorentzian scalar product between two 4D vectors V and V' is: $V•V' = c^2 t.t' - (xx' + yy' + zz')$. We can easily check that $ds^2(V) = V•V = c^2 t^2 - x^2 - y^2 - z^2$.

If the frame is not orthonormal, the Lorentzian scalar product can be obtained by the general relation between the norm^2 and the corresponding scalar product:

$$V•V' = [ds^2(V + V') - ds^2(V) - ds^2(V')]/2.$$

This relation also shows that the Lorentzian scalar is intrinsic since the Lorentzian norm is.

Physically, the Lorentzian distance definition between an event A and an event B corresponds to the time duration for an inertial traveler to go from A to B, and measured with his proper time (cf. Section 3.2.4.1.2). (This only concerns time-like intervals.) If the Lorentzian distance between A and B is null, it means that there can only be photons going from A to B.

We also saw the surprising triangle inequality, $ds^2(AB) > ds^2(AC) + ds^2(CB)$, with its amazing consequence: when an inertial traveler goes from an event A to an event B, he chooses the path which maximizes his travel duration as measured with his proper time (cf. Langevin's paradox), and this path is a geodesic.

Another difference with the classical distance is the following: all events that are at equal Lorentzian distance to the same event form a hyperbolic surface, and not a sphere (cf. figure 3.8).

6.3.3 Relationship Between Distance and Chrono-Geometry

Let's pursue our analogy with a block of jelly: we now add a fourth dimension for the time. Before any distortion, we have a 4D orthonormal frame where we can draw a network of small cubes, but now we must imagine a 4th dimension for the time. The presence of mass produces distortions, which transform the cubes into small 4D parallelepipeds. Our traveler passes through a succession of such parallelepipeds, and inside each one, he can benefit from the two following properties:

- Linear approximation: the sides of the small parallelepiped are straight lines. We can then define a frame, denoted by K', with its unit basis vectors matching with the sides of the parallelepiped, similarly as what was previously done with our 3D block of jelly.

- The principle of equivalence: laws of Special Relativity apply for a free-falling traveler inside the parallelepiped. Thus inside each parallelepiped, his trajectory is inertial: along a straight line and with constant velocity.

K' thus is a Euclidean tangent frame where the laws of Special Relativity apply. Our free-falling traveler enters one very small 4D parallelepiped at the event M and exits at the event N.

Let's find the relationship between the distance $ds^2(MN)$ and the chrono-geometry of the small 4D parallelepiped:

MN being very small, we will call it dM with its coordinates in the tangent frame K' being: (cdt', dx', dy', dz'). We want to calculate the distance of dM, meaning $ds^2(dM)$.

Unfortunately, the formula $ds^2(dM) = c^2 dt'^2 - (dx'^2 + dy'^2 + dz'^2)$ cannot apply because K' is not orthonormal: we indeed chose the axes of K' to be the sides of the parallelepiped, and we did so in order to express the link between the chrono-geometrical distortions and the distance (as we previously did with 2D and 3D surfaces). We will see that it also works for 4D:

The norm of dM can be calculated applying the same principles as with our jelly, but adapting to 4D our previous relations (6.10), by replacing 3D vectors with 4D space-time vectors, and replacing the classical scalar product with the Lorentzian one. We thus obtain:

$$ds^2(dM) = dM \bullet dM = (c.dt'\vec{f'} + dx'\vec{i'} + dy'\vec{j'} + dz'\vec{k'}) \bullet (c.dt'\vec{f'} + dx'\vec{i'} + dy'\vec{j'} + dz'\vec{k'})$$
$$(6.12)$$

where $\vec{f'}$ is the unit vector of K' corresponding to the time axis; and as previously, $\vec{i'}$, $\vec{j'}$, $\vec{k'}$ are the unit spatial vectors which match with the sides of the parallelepiped.

When expanding equation 6.12, we obtain a quadratic form of the coordinates of dM in the tangent frame:

$$ds^2(\mathbf{dM}) = g_{\mu\nu}dx'^{\mu}dx'^{\nu} \blacksquare \qquad (6.13)$$

or using Einstein's compact notation: $ds^2(\mathbf{dM}) = g_{\mu\nu}dx'^{\mu}dx'^{\nu}$ ∎

The terms cdt', dx', dy', dz' are, respectively, denoted by: dx^0, dx^1, dx^2, dx^3. The letters μ and ν are indices taking successively the 4 values (0, 1, 2, 3) corresponding to the 4 dimensions of the space-time universe. Similarly, the unit basis vectors $\vec{f}', \vec{i}', \vec{j}', \vec{k}'$ are, respectively, denoted by: $(\vec{i_0}, \vec{i_1}, \vec{i_2}, \vec{i_3})$.

Each $g_{\mu\nu}$ is the Lorentzian scalar product between the corresponding unit basis vectors:

$$g_{\mu\nu} = \vec{i}'_{\mu} \bullet \vec{i}'_{\nu}. \qquad (6.14)$$

There are 16 $g_{\mu\nu}$ coefficients, but as the Lorentzian scalar product is commutative, $g_{\mu\nu} = g_{\nu\mu}$, there are only 10 different coefficients forming a symmetric 4×4 matrix. Thus, the characterization of space-time distortions requires 10 coefficients at every event.

Remark 6.11: The matrix formed with the $g_{\mu\nu}$ coefficients is a rank 2 tensor which is twice covariant, as shown in Volume II Section 4.1.2. It is called the **"metric tensor"** since it enables us to calculate the distance $ds^2(\mathbf{MN})$ in the neighborhood of the event M.

Similarly, as with 3D flexible volumes, the $g_{\mu\nu}$ coefficients and their variations (derivatives) around the event M reflect the chrono-geometry of this region, and thus its deformations. This can be seen the following way: the Lorentzian scalar product being invariant, it can be calculated in any frame, in particular in the orthonormal frame that existed before the distortions due to the mass. In this frame, denoted by Ko, the unit basis vectors are denoted by: $\vec{i_t}, \vec{i_x}, \vec{i_y}, \vec{i_z}$.

The coordinates of the vector $\vec{i_{\mu}}$ in this basis are denoted by $(i_{\mu t}, i_{\mu x}, i_{\mu y}, i_{\mu z})$. The Lorentzian scalar product between $\vec{i_{\mu}}$ and $\vec{i_{\nu}}$ is then:

$$g_{\mu\nu} = \vec{i_{\mu}} \bullet \vec{i_{\nu}} = c^2 i'_{\mu t} i'_{\nu t} - (i'_{\mu x}.i'_{\nu x} + i'_{\mu y}.i'_{\nu y} + i'_{\mu z}.i'_{\nu y}). \qquad (6.15)$$

The last term is the classical scalar product between the 3D vectors made of the spatial parts of $\vec{i_{\mu}}$ and $\vec{i_{\nu}}$. Let's denote these vectors by $\vec{i_{\mu S}}$ and $\vec{i_{\nu S}}$; we have:

$$\vec{i_{\mu S}} = i_{\mu x}.\vec{i_x} + i_{\mu y}.\vec{i_y} + i_{\mu z}.\vec{i_z}$$

and the same for $\vec{i_{\nu S}}$. The relation 6.15 becomes:

$$g_{\mu\nu} = \vec{i_{\mu}} \bullet \vec{i_{\nu}} = c^2 i'_{\mu t} i'_{\nu t} - \vec{i_{\mu S}}.\vec{i_{\nu S}} = c^2 i'_{\mu t} i'_{\nu t} - i_{\mu S} i_{\nu S} \cos(\vec{i_{\mu S}}, \vec{i_{\nu S}}).$$

This relation shows the link between the chrono-geometry and the $g_{\mu\nu}$ coefficients. In particular, if the small parallelepiped hasn't been distorted, its sides remain perpendicular one to another; hence, the cosine between two different sides is null.

Having seen how to calculate the (Lorentzian) distance between any couple of close events, we can calculate the distance along any trajectory between two distant events A and B: by adding up (integrating) the small distances in the succession of small 4D parallelepipeds that are crossed. A case of special interest is the trajectory of an inertial (free-falling) object: the geodesic.

6.3.4 The Geodesic: Main Principles

We saw that a free-falling object has its Momentum constant, meaning $P = m.dM/d\tau = cst$. This implies that the vector dM, which is along the tangent to the object trajectory at the event M, is always parallel to the same direction, and this is a characteristic property of the geodesic.

Then according to the other characteristic of the geodesic, the Lorentzian distance between any couple of events is maximized. Note that a free-falling object is only driven by the immediate chrono-geometry in its vicinity; hence, this may lead him to choose a path between two distant events A and B, which is not the longest one, depending on the surface curvature. Nevertheless, it is still a geodesic, meaning a trajectory which *locally* maximizes the distance covered. Here, locally means that a small change in the coordinates of the events along this trajectory leads to a slightly smaller distance AB.

The mass of a free-falling object remains constant since it is the norm of its Momentum, and the latter remains constant. Then, if we denote by S the object proper velocity, $S = dM/d\tau$, the object trajectory is determined by: $dS/d\tau = 0$. We can first see that the mass of a free-falling object has no effect on its trajectory. (We neglect the friction effects with the atmosphere.) Its trajectory is only dictated by its initial velocity and the chrono-geometry of his region.

The calculation of the geodesic trajectory requires mathematical notions, which are presented in Volume II Section 4.2. The difficulty stems from the fact that the proper velocity vector S, which must remain constant, is in a different tangent space at the event M (time τ) and at the event $M + dM$ (time $\tau + d\tau$) due to the chrono-geometrical distortions around M. The variation (derivative) of this vector S must then take into account the chrono-geometry around M, and this is achieved thanks to a mathematical tool called "covariant derivative" based on the "affine connection". This tool involves the different $g_{\mu\nu}$ and their derivatives, reflecting the chrono-geometry around M. The covariant derivative $dS/d\tau$ must constantly be null; hence, the proper velocity vector S incurs what is called a "parallel transport" along the object trajectory.

Remark 6.12: Photons' trajectories are geodesics of distance null, called "null geodesic", corresponding to successions of small paths of distance null ($ds^2 = 0$). The constancy of the photon Momentum implies that its covariant derivative is null, but if the space is distorted, this implies that the components of its Momentum change, which is consistent with the result previously seen that a photon gains energy when falling toward the great mass, and conversely loses energy when going away from it.

6.4 The Equations of General Relativity

6.4.1 Main Principles Leading to the Equations of General Relativity

The main idea is that our space-time universe has some elasticity, and that the presence of masses (energy) generates chrono-geometrical distortions which propagate at the speed of light.

A similar phenomenon occurred in classical physics, and this will give us the clue: the relationship between the distortions within an elastic medium and the stresses applied to it.

When we exert some stresses at a certain part of a medium having some elasticity (such as a block of jelly), we will not only distort this part, but this distorted part will in turn exert stresses on its neighboring parts which will distort them, and so on. Distortions will thus propagate in the medium while progressively fading.

The greater the medium elasticity, the more important the deformations. However, beyond some intensity, there can be a breaking point. We will assume that we are far from this breaking point, and that the function giving the deformations knowing the stresses is linear.

> The simplest example is the spring for which we have: $d = \kappa.t$, with d being the length of the spring deformation, κ its elasticity and t the tension (force) applied.

Consider again deformations concerning a volume such as a block of jelly. We saw in Section 6.3.2 that, in a small parallelepiped around any point, the deformations can be characterized by six coefficients forming a deformation matrix. The mechanics of continuous media showed that the constraints (tensions, stress) applied on a small volume inside a medium could be characterized by six parameters, with a similar form as the parameters characterizing the deformations of a small parallelepiped. These parameters were expressed by a constraint matrix, called a tensor and denoted by **T**. (The origin of the word tensor comes from there.). This stress tensor **T** has the same mathematical structure as the deformation matrix; hence, the relationship between the stress applied to an elastic media and its subsequent distortions could be expressed in a simple and linear way:

$$\mathbf{D} = \kappa.\mathbf{T}$$

where κ is the elasticity coefficient of the structure, **T** is the tensor characterizing the stresses, and **D** is the tensor characterizing the deformations (the same as in Section 6.3.2).

Let's then adapt these principles to our space-time universe even though there is no real jelly, as gravitation doesn't need any medium to propagate.

We saw that the deformations of our space-time universe can be modeled with the metric tensor $(g_{\mu\nu})$, which has 10 different coefficients (out of 16).

Einstein found that the stress-energy tensor **T** also has ten different coefficients: one reflecting the density of mass-energy per unit of volume (time part of the Momentum); three for the density of impulse (spatial part of the Momentum), and the six remaining for measuring the stresses on this small volume, similarly as for the classical ones in our block of jelly (cf. more in Volume II Section 4.1.2.5). These stresses on the small volume take into account the propagation of the distortions.

Hence, Einstein wanted the law of General Relativity to be: **D**=κ.**T**, meaning that the deformations are proportional to the energy-stress, with κ being the elasticity of our space-time universe.

Remark 6.13: The notion of mass-energy involved in General Relativity is the one of Special Relativity. It encompasses all sorts of energy, matter and radiation: all generate gravitational effects.

But the devil is in the details: an important mathematical difficulty arose because the principle of equivalence (gravity ⇔ acceleration) is only absolutely valid at one point (cf. the tidal effect, Section 6.1.2). In the neighborhood of this point, there is a small remaining gravitation effect. This phenomenon could be taken into account by the deformation tensor of Riemann-Christoffel, which has 20 parameters (after the reduction due to symmetries), double the number of the previous deformation and stress tensors.

Einstein succeeded in reducing to ten the number of coefficients of the final deformation tensor, called Einstein's tensor and denoted by **D(g)**, which is proportional to the stress-energy tensor **T**, and which characterizes most deformations. He did so by setting the two following principles (postulates):

- All physical laws must take the same form whatever the system of coordinates used, which he called the principle of General Covariance, and also General Relativity.
- The conservation of the Momentum (energy-momentum tensor).

Einstein's famous General Relativity equation thus reads: **D(g)** = κ.**T** .

This relation is not as simple as it may seem: it expresses ten equations relating the space-time deformations to their causes (presence of mass-energy). These are non-linear second-order differential equations. The coefficients of Einstein's tensor are composed of the different $g_{\mu\nu}$ and their derivatives of the first and second orders. (It is thus quite complex in the general case.)

The elasticity coefficient κ could be calculated by requiring that this General Relativity relation gives the classical Newtonian law as an approximation, $\dfrac{Gmm'}{r^2}$, with G being the classical gravitation coefficient: $G = 6.674\ 30 \times 10^{-11} \text{m}^3\text{kg}^{-1}\text{s}^{-2}$.

The result is: $\kappa = 8\pi G/c^4$,

which gives: $\kappa = 2.10^{-48}$ with the commonly used units by scientists, i.e., distances in centimeter, weights in gram, time in second. This result shows that the elasticity of our space-time universe is extremely little: it is extremely rigid, which explains why we did not notice that it was deformable.

The equations of General Relativity have been confronted with many different experiments and observations, and all gave extremely accurate results, in the contexts of both low gravity and extreme ones as with black holes and even gravitational waves (cf. Sections 7.2 and 7.3).

Remarkably, the Newtonian approximation appears to be very good, the difference with that of General Relativity being in the range of 10^{-9} on the Earth's surface and 10^{-6} on the Sun's surface. This difference increases near to the center of great masses, and is huge in case of very strong gravity such as black holes.

6.4.2 Confirmations of General Relativity: Mercury's Perihelion and the Bending of Light

6.4.2.1 First Confirmation: Solution to the Mercury Perihelion Enigma

Einstein issued a very first confirmation of General Relativity in November 1915 with the solution of the enigma of Mercury perihelion: Mercury was known as the only planet which did not strictly follow Newton's laws: its trajectory is an ellipse which is continuously deviating from its long axis, as shown in the picture below (which greatly amplifies the phenomenon). The magnitude of this precession is greater by a small amount than what was predicted by the Newtonian theory, i.e., 43 seconds of arc per century, but this excess was recurrent over more than 50 years of observations, and remained mysterious (Figure 6.9).

When Einstein saw that his calculations gave the correct deviation angle for Mercury with a precision better than 1%, he remained "speechless for several days with excitement[3]". These calculations are quite complex; hence, we won't present them but only a qualitative description.

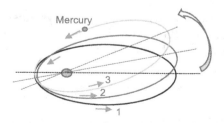

FIGURE 6.9
Precession of Mercury's perihelion.

A planet, as any object, chooses the trajectory (worldline) which maximizes its trajectory duration measured with its proper time. Einstein understood as soon as 1912 that it implied that the planet must coil itself around the Sun so as to maximize its worldline length. Note that this effect exists for other planets but is less important, Mercury being the closest planet to the Sun. For example, in the case of the Earth, it is 5 seconds of arc.

6.4.2.2 Schwarzschild's Solution and the Bending of Light

The Schwarzschild solution is presented in Volume II Section 5.2. A general overview is given hereafter:

Just 1 month after Einstein published his General Relativity theory, the mathematician and astronomer K. Schwarzschild found a solution of Einstein's equations in the context of a massive spherical body, such as a planet or the Sun. This solution enabled astronomers to calculate the deviation of light near a great mass such as the Sun, which could then be experimentally checked in 1919.

This famous observation was indeed conducted by the British astronomer A. Eddington during a total eclipse of the Sun in Sao Tome-and-Principe (Figure 6.10).

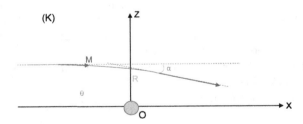

FIGURE 6.10
The bending of light near a great mass.

In order to have a maximum deflection, the rays must pass as close as possible to the Sun; therefore, their observations were made when the rays from the Sun were stopped by the Moon. The prediction of General Relativity was:

$\alpha \approx \dfrac{4GM}{c^2 R}$, which yields a deviation angle of 1.75 second of arc.

The first series of observations gave: 1.98 ± 0.16 second of arch; a second one:

1.61 ± 0.4. These provided a very important confirmation of Einstein's General Relativity, showing in particular that our universe is non-Euclidean but distorted by the presence of a great mass.

Furthermore, the Schwarzschild solution opened the possibility of the existence of black holes, and this will be the subject of Section 7.2.

6.5 Questions and Problems

Problem: 6.1: Tidal effect from the Sun and from the Moon

Q1: Compare the attraction from the Sun and that from the Moon on the Earth surface, using the Newtonian approximation (which is very good in these cases). Which is the greater and by how much? You may use the values that are in Section 10.

Q2: How do you explain that the tidal effect from the Moon is more important than from the Sun? Calculate the gravity difference between the point on Earth that is the closest to the Moon, and the one which is the furthest. Same questions for the Sun. What do you conclude?

Problem 6.2: Why can't we see light bending?

Q1: When we throw an object horizontally, it starts moving horizontally and then we generally can see it falling. Why is it not the case for light?

Q2: A light beam is sent horizontally on Earth. We measure the altitude of the beam at its emission location, and when it has covered a distance of 1 km. We assume that gravity is the same all along the beam trajectory, with $g=9.81$ m/s². What is the difference of the beam altitude between its emission and after 10 km?

Problem: 6.3: Amazing Effects of Travels in the Cosmos – Bob and Alice Case Study 2

This is a continuation of the previous Problem 2.3. We will now consider phenomena that are due to acceleration, and then to gravity. Qualitative answers will suffice.

Bob and Alice are twins; Bob is undertaking a travel in the cosmos, whereas Alice stays on Earth. Both set their clocks at zero when Bob leaves Alice. Bob's rocket goes at a very fast speed of 0.9 c, and then comes back to Earth to meet with Alice.

Q1: When Bob returns and meets Alice again, will he be older or younger than her? We assume Bob didn't stay near a massive planet during his journey. Besides, we neglect the effect of terrestrial gravity on the time (incurred by Alice).

Q2: Same as Question 1, but now Bob stops for few years in a planet which is more massive than the Earth, and then comes back to see Alice. Who will be the younger?

Q3: Same as Question 1, but we do not neglect the effect of the terrestrial gravity on the time incurred by Alice. Does this effect alone make Alice younger or older than Bob at Bob's return?

Q4A: Bob stops his rocket when he is at the altitude 100,000 km from Earth, and he is not near a planet. He phones to Alice: Does Alice's voice sounds to him like usual, slightly more treble or more bass?

Q4B: Does Bob's voice sounds to Alice unchanged, or slightly more treble or more bass?

Problem: 6.4: The Clock in a Geostationary Satellite

A satellite is orbiting the Earth at an altitude of 36,000 km above the ocean level. It has a perfect clock, identical to one made on Earth.

Q1: For an observer inside the satellite, is this clock beating at the same pace as when it was on the Earth surface?

Q2A: For an observer on Earth, is the satellite clock seen as fast or slow? By how much during 24 hours? You may use the Newtonian gravitation expression.

Q2B: Express this time difference during 24 hours in distance covered by light, knowing that $c=30$ cm/ns. This gives the range of error that the GPS would have if Relativity was ignored.

Q3: This satellite emits a signal to the Earth at the frequency f_s. Is it received with the same frequency on Earth, and if not, what is the relation between the two?

Problem 6.5: Curved surface: the Sum of the Angles of a Triangle

Imagine that you are on a curved surface which is a sphere. Take two points A and B on the equator, and the North Pole is denoted by N. Consider the triangle ABN formed by the portion of equator between A and B, the portion of meridian between A and N, and the portion of meridian between B and N. The angle at the North Pole between the meridians NA and NB is denoted by α.

Q1: Is the sum of the three angles of this triangle equal to 180°? Please comment.

Q2: Calculate the sum of the three angles, denoted by α, β, γ as a function of the radius of the sphere, R, and the surface area of ABN, denoted by S.
Show that: $\alpha + \beta + \gamma = \pi + S/R^2$.

Problem 6-5: The Angles
on a Curved Surface

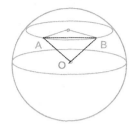

Problem 6-6: The
Geodesic on a Sphere

FIGURE 6.11
Angles and Geodesic on a Curved Surface.

Problem 6.6: Curved surface: What is the Shortest Path?

Consider two points A and B on Earth that are on the same latitude $\theta=45\%$, and the difference between their longitudes is $\varphi=90°$. You want to go from A to B.

Q1: Is the path along the latitude 45° the shortest?

Q2: If not, what is it? Determine the length difference in percentage. The Earth radius is denoted by R, and its center by O.

Notes

1 Cf. https://einsteinpapers.press.princeton.edu/vol7-trans/152 – Volume 7: The Berlin Years: Writings, 1918–1921 (English translation supplement), page 136.
2 This is also true in Relativity; cf. Volume II § 3.3.5.
3 Cf. Einstein to Ehrenfest, January 1916, Seelig, page 156. Also quoted by Ronald W. Clarck in Einstein, *The Life and Times*, page 206.

7
Cosmological Consequences

Introduction

Three important cosmological predictions of General Relativity will be presented: the Big Bang theory, black holes and gravitational waves. In its early years, these predictions were hurting the theory's reception, particularly as they weren't supported by any observational evidence. Hence, the main steps that led to their recognition will be depicted, remaining at the level of the general principles.

7.1 The Big Bang Theory

In 1917, Einstein completed his General Relativity equation with the addition of a constant term called the cosmological constant. He did so because he realized that his equation implies that the space is a function of time: either the universe expands or it contracts. In the early twentieth century, the Milky Way was the entire known universe, and its stars were considered to be fixed. Hence, such universal expansion (or contraction) was not conceivable, neither by Einstein nor by the scientific community. He then realized that his equation could accommodate an additional constant term which, with a special value, yields a static universe. Hence, he proposed it.

In 1922, a young Russian mathematician, Aleksandr Friedmann, showed that this cosmological constant made the universe unstable, and proposed an expanding universe model with the advantage of explaining why the universe doesn't collapse from the effects of gravitation: the Big Bang theory was born (even if the term Big Bang was coined 20 years later). Indeed, when going back in time, the universe size would tend to zero volume, while its density would be extremely high, assuming that mass-energy remained constant. This extremely hot area would then have expanded extremely rapidly, hence the name "Big Bang". In the beginning, Einstein was strongly opposed to this interpretation. So were many scientists of the time, some being perplexed about the mysterious initial state which would open the way to some divine explanation.

DOI: 10.1201/9781003201335-7

Note that the classical Newtonian gravitation theory also predicted that the universe would collapse if it had a center, such as the Sun. Therefore, Newton believed that the universe was infinite, so that at each point, the masses would be equally distributed in all directions.

In 1925, the American astronomer Edwin Hubble showed the existence of other galaxies and noticed a linear relation between the distance of a galaxy and its redshift. However, Hubble didn't interpret this relation as a mark of the expansion of the universe. This interpretation can be credited to Belgium astronomer and scientist Georges Lemaître who proposed it in 1927, confirming what had been predicted by Friedmann – who had died in 1925. This redshift is different from the one due to the Doppler effect, and even from the gravitational one: it is a cosmological redshift due to the swelling of the universe between very distant objects. When a photon travels a long distance, its wavelength is dilated because it is affected by the expansion of the universe, which also means a drop in frequency. Locally, however, object sizes are not affected by the expansion of the universe due to local interaction forces, including gravitation. Similarly, the size of a galaxy is much less affected than expected due to the gravitational effects between its masses.

In 1964, the discovery of the Cosmic Microwave Background brought an important confirmation of the Big Bang theory: two radio astronomers, Penzias and Wilson, discovered by chance the existence of a permanent low-level electromagnetic signal on Earth and presumably in the universe, whose spectrum perfectly corresponds to the emission by a black body at the temperature of 2.73 K. Surprisingly enough, it is extremely stable and isotropic, as shown in the picture below, its fluctuations being less than 10^{-5} degree K (Figure 7.1).

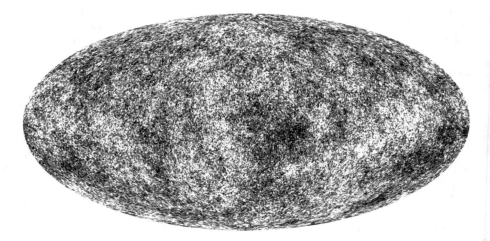

FIGURE 7.1
Cosmic Microwave Background. The temperature difference from the average temperature is + 200 μK for the white dots and –200 μK for the black dots. (NASA/WMAP Science Team-ESA-Vincent Pajot.)

This discovery actually matched a prediction made in 1948 by G. Gamow, R. Alpher, R. Herman and J. Peebles, based on Lemaître's work: according to the Big Bang model, the universe was so dense and hot in its initial phase that atoms were ionized: it consisted of a plasma of particles (mainly protons, electrons, photons, etc.). During the first 380,000 years, photons were captured by electrons very rapidly after being emitted. Then, as the universe expanded very quickly, it cooled down, and when its temperature reached the 3,000 K range started the era of recombination whereby electrons combined with protons, forming atoms. Consequently, many photons became free to move along long distances without either being absorbed or encountering obstacles, and these are the photons that we observe today with this low signal forming the Cosmic Microwave Background. Having been emitted 13.8 billion years ago, they have crossed a huge distance, estimated at 45.6 billion light-years, which is greater than 13.8 (the age of the universe) due to the considerable expansion of the universe during their travel. It is indeed estimated that the universe has expanded by a factor of 1,090 since the recombination era. This very significant expansion also tells us that these photons were extremely energetic at their emission, corresponding to the temperature of 2,790 K. Still today, despite their extremely low energy of $\approx 6.3 \times 10^{-4}$ eV, they represent a huge energy due to the fact they are omnipresent: they account for 90% of the total radiation energy of the universe (Figure 7.2).

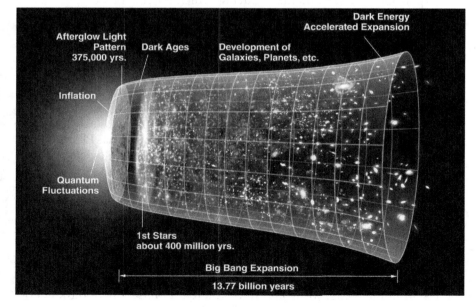

FIGURE 7.2
Timeline of the Universe. (NASA/WMAP Science Team.)

Some 100 million years after the Big Bang, atoms gathered into clusters due to gravitation, forming the first stars and galaxies. The extremely isotropic character of the Cosmic Microwave Background is a confirmation of the postulate, strongly supported by Einstein, that the universe is homogeneous and isotropic. The formation of structures such as clusters of galaxies stems from very small perturbations which increased over time due to gravitation, their origin being probably quantum-mechanical.

Besides, Relativity has revealed an equivalence between time and distance; hence, one may question whether the expansion of distance is compatible with that principle. Actually, there is also a cosmological time dilatation: any phenomenon appears to be slower to a very distant observer.

The current expansion is in the range of 5%–10% in 10 billion years, much lower than it was in the early stages. The observable size of the universe is 45.6 billion light-years, that is, the distance to the furthest objects from which we receive information: these are the photons of the Cosmic Microwave Background. However, the universe is wider than its observable size, by at least a few tens of billions of light-years. We don't know whether the universe is finite (S. Hawking believes so) or infinite (R. Penrose). Receiving information from further than 45.6 billion light-years means that information was emitted before the recombination era. Hopefully, we will be able to receive such information, thanks to gravitational waves detectors (cf. Section 7.3).

It is assumed that before the recombination era, from 10^{-35} seconds after the Big Bang till 10^{-32} seconds, the universe has incurred a period of extremely fast expansion, called "inflation", during which its size has enlarged by a factor of 10^{23} ! This hypothesis explains in particular why the Cosmic Microwave Background is so homogeneous despite its size. Such expansion was at a greater speed than c, but it doesn't contradict Relativity since it concerns space itself and not this local speed of objects. Information exchange between distant objects is still impossible at a greater speed than c.

The state of the universe at time zero is still mysterious: it is a singularity in which the laws of General Relativity do not apply since they give space an infinite curvature as shown by S. Hawking and R. Penrose. Immediately after time zero, our laws still don't apply, until 5.39×10^{-44} seconds, which is Planck's limit.

In 1931, Einstein regretted having introduced his cosmological constant and even viewed it as his "biggest blunder[1]". However, recent developments have rehabilitated this constant after the observations in 1998 that the expansion of the universe was in an accelerating phase. That was against cosmologists' predictions since during the first half of its lifetime, the expansion of the universe was decelerating. The current explanation for this change lies in the cosmological constant which has repulsive gravitational effects. Conveniently, quantum mechanics explains that this cosmological constant represents the energy of the vacuum, dubbed "dark energy". Thus, over the last 4 billion years, we would have entered a phase where dark energy is dominant and now accounts for 68% of the total energy in the universe despite its extremely low energy density. A positive effect of this acceleration is that it avoids the "Big Crunch" that would ultimately result from an ever-decelerating expansion. However, the cosmological constant is not the only explanation and remains an important subject of research.

In addition, the acceleration of expansion is not the only major unexplained phenomenon: the speeds of many stars are greater than expected from the Keplerian rotation law[2]; hence, the existence of a "dark halo" surrounding galaxies and made of an unknown matter, called "dark matter," was suggested: it would have an attractive gravitational effect like any mass, but it would not interact with photons (which explains why we cannot see it). This dark matter would account for up to 27% of the total energy in the universe, and so the part which is known accounts for only 5% of the total!

7.2 Black Holes: The Long and Chaotic Road That Led to Their Recognition and Discovery

Without Relativity, we wouldn't know of the existence of black holes even though they are the most massive bodies in the universe. However, their discovery was a long and chaotic process as we will see:

The possible existence of black holes was anticipated in the 18th century by J. Mitchell and P.S. Laplace. Their reasoning was the following: light was thought to be made of very small corpuscles having a mass which is not null. Hence, the famous Newtonian laws imply that for these light corpuscles to escape from the gravitation field of a great mass, they must have a velocity greater than the critical "liberation speed[3]". Mitchell and Laplace were able to calculate the liberation speed for the Earth and found that it was 11 km/s, that is, much lower than the speed of light that was known at that time. However, J. Mitchell calculated that if a celestial body had the mass of the sun, and a circumference of 18.5 km only, then the liberation speed would equal the speed of light. As, for denser stars, light could not escape their gravitation field, observers from far away could not see any light coming from these stars; hence, such stars were called "black stars".

Mitchell and Laplace believed that there could be many such black stars in the universe, and the scientific community did not object. In the early 1800s however, T. Young and C. Huygens showed that light was a wave. That completely ruined Mitchell's and Laplace's reasoning, and the potential existence of black stars was discarded.

Much later, Relativity showed that light follows null geodesics in the space-time universe, which are not straight lines near great masses due to chrono-geometry distortions caused by these masses. In 1916, Schwarzschild described a solution to the General Relativity equations which enabled calculation of geodesics in the neighborhood of massive spheres. A surprising result was that if a massive sphere is smaller than its Schwarzschild radius,

$r_s = \dfrac{2GM}{c^2}$, nothing including light can escape from any point inside a sphere of Schwarzschild radius r_s: space-time distortions are so great that everything falls to the center.

However, such a possibility was very unlikely to occur in the universe because the radius of all known planets and stars exceeded by far their Schwarzschild radius. Indeed, for the Earth, r_s = 8.9 mm; for the Sun: 2.954 km. A star that would be smaller than its Schwarzschild radius would be denser than anything known; hence the question: could such an extremely great and dense mass exist in the universe? The scientific community strongly rejected this possibility until the 1960s, though with a few exceptions, in particular R. Oppenheimer and from the beginning S. Chandrasekhar.

The bodies with the highest density of matter that were known in the 1920s were the "white dwarfs". These stars are much smaller than most stars, and with much lower brightness. The closest one that was observed (by telescope) was Sirius B, which was orbiting around the brightest star, Sirius. This white dwarf Sirius B appeared to have a mass of 0.85 solar masses and a circumference of 118,000 km, meaning a volume 51,000 times less than the Sun; hence, a huge density: 61,000 times that of water! This was considered absurd until 1925 when W.S. Adams confirmed this extreme density by measuring the significant gravitational redshift of the light emitted by Sirius B.

White dwarfs raised a serious theoretical issue: how can matter overcome such a huge pressure due to gravity, especially in their cores? The classical explanation was that the random movements of atoms, which account for the heat, generate a pressure which overcomes gravity. Then as Sirius B radiates, it loses energy and so its radiating pressure diminishes, causing the external layers to move toward the center, creating a compression which in turn generates heat. Thus as Sirius B radiates, new equilibria are met whereby its volume diminishes and its density increases. The logical end of this process is the creation of a black hole, but A. Eddington, who was the most renowned astronomer at that time, strongly rejected this possibility: in his mind, nature must provide mechanisms forbidding the creation of black holes (even if the name "black hole" wasn't invented yet), but he couldn't tell what they were. He rejected the possibility that the resistance to compression present in usual matter such as stone – which is due to the repulsion between adjacent atoms – could apply to very high pressures; he was also wrong.

The solution to this enigma came in 1930 from a young Indian student, S. Chandrasekhar, who based his reasoning on a 1926 article on dense matter by R. Fowler: when matter is compressed 10,000 times more than ordinary stone, the space where the electrons move is extremely reduced, and this causes the electrons to drastically increase their speeds. This stems from the principle of wave-particle duality in quantum mechanics: as the space available for the electron shrinks, its wavelength must also shrink, but lower wavelength means higher energy, and so the electrons increase their speeds; consequently, the matter can exert a higher pressure enabling it to resist to higher pressures from gravity. Moreover, there is another phenomenon working in the same direction: the Pauli exclusion principle implies that two electrons cannot be simultaneously in the same state and same orbital, which forces electrons to create a vital space around them. Both phenomena result in what was called "degeneracy pressure", which enables matter to resist up to extremely high pressures.

Chandrasekhar made the corresponding calculations for Sirius B: the result was that electrons moved with extremely high velocity in the center: 0.57 c! He realized that the laws of Relativity should be applied, which hadn't been done before, and he found out that the electron' energy increase, due to its space reduction, not only translated into higher speed but also into higher *inertia*. (Note that inertia is: m.γ.) Thus, the faster the electron, the less incremental rise in speed it will incur for the same rise in energy. Consequently, Chandrasekhar showed that degeneracy pressure cannot offset gravity pressure if the white dwarf mass exceeds 1.4 solar masses.

White dwarfs were considered an advanced stage of stars, the mechanism leading to their formation being the following: a star like the sun is extremely hot, and its heat exerts a pressure which prevents the particles around the core from falling into it due to gravity. This phenomenon also explains the relatively low density of the sun, three times lower than that of the Earth. The heat is due to hydrogen thermonuclear reactions in the core of the sun, hence the question: what happens when all the fuel (hydrogen) has been consumed? The pressure of radiation will disappear, and so all the matter will fall toward the center, generating extremely high pressures. However, Chandrasekhar's conclusion was that if the star mass exceeded 1.4 the sun mass, the repulsion between adjacent atoms could not prevent a further crunch; he then concluded: "One is left speculating on other possibilities. " The logical avenue was the black hole, but for the scientific community, this was still unacceptable.

The next important finding on the way to black holes occurred in 1933 when astronomer F. Zwicky issued two bold hypotheses: the existence of neutron stars and subsequently the explanation of supernovae based on neutron stars. The neutron itself was found in 1931 by J. Chadwick who showed that the atom nucleus is made of two building blocks: protons and neutrons. Zwicky had a passion for supernovae, which are the brightest stars, equivalent in brightness to a whole galaxy. However, they only last for several weeks or months. Zwicky stated that a supernova actually was the huge explosion of a very massive star, after it had consumed all its fuel, and that the degeneracy pressure of its electrons couldn't stop its implosion. Then atoms were destroyed, protons and electrons combined forming neutrons, so that finally only a very dense neutron star remained. Zwicky's reputation was not very high as a physicist; hence, his statement was deemed too speculative and most scientists didn't even consider it. Nevertheless, one of the most renowned scientists worldwide, the Russian L. Landau, took up Zwicky's neutron star idea, and published an article in 1937 stating that the Sun core is made of neutrons. This actually was a desperate move on his part in order to avoid the Gulag (with a high risk of death), hoping that the Russian authorities would shy away from targeting a world's top scientist. Unfortunately, Landau was sent to the Gulag and his thesis soon proved to be wrong. But fortunately, his article triggered the attention of other top scientists, in particular R. Oppenheimer who was a highly renowned professor at Berkeley. Oppenheimer showed in 1939 that the maximum mass of a neutron star was a few times the mass of the sun. The principle of his reasoning was similar to Chandrasekhar's, but applied on neutrons instead of electrons. Neutrons also follow the laws of quantum mechanics, including the Pauli exclusion principle, and so when their vital space is reduced, they incur a degenerate behavior enabling them to resist higher pressures, but up to a certain point. Additionally, neutrons are subject to a nuclear force which is repulsive and further increases the resistance to compression, due to the state of the gas in which the neutrons move freely. As the pressure increases, the core becomes a solid "crystal" of neutrons.

Consequently, for Oppenheimer, the destiny of stars that were substantially heavier than the sun was to become black holes: once the fuel was consumed, nothing could stop an implosion; nothing could withstand the huge pressure from gravity, leading to extreme density in the center. However, the majority of scientists remained with Eddington's position until the sixties: they believed that there exist unknown mechanisms whereby very massive stars eject enough matter, as in the case of the supernovae.

During World War II, like most other top scientists, Oppenheimer was asked to work on advanced armament and unfortunately dropped his research on black holes to become the father of the atomic bomb. In the fifties, J. Wheeler, a top scientist on nuclear fission and later on fusion, took up the studies of the destiny of massive stars. He confirmed and completed the calculations of both Chandrasekhar and Oppenheimer, but rejected their common conclusion regarding the possibility of black holes, stating like Eddington that nature must provide mechanisms preventing them. Oppenheimer remained confident about his previous conclusion that black holes were the destiny of the biggest stars. Both undertook to better characterize the implosion of great masses, capitalizing on the important work that had been done on the atomic bomb. In addition, D. Finkelstein brought in a major contribution by inventing a new way of representing space-time, which is better adapted to black holes than Schwarzschild's. In the early sixties, they both concluded that the implosion of such great stars must result in a black hole! Wheeler even became a strong supporter of the theory, and he coined the term "black hole" for what was previously called the "Schwarzschild singularity".

There was widespread use of General Relativity although the theory had only been validated in the solar system where space-time distortions were limited. In the context of black holes, the results of General Relativity are very different from Newton's: in particular, gravity inside the black hole is much more intense than Newton's prediction, even leading to a singularity in the center where space-time distortions tend to infinity. This was shown by R. Penrose with an original mathematical reasoning based on topology. However, infinity being unacceptable in physics, General Relativity was combined with quantum mechanics, forming quantum gravity, with the conclusion that this singularity takes place in an extremely small region – in the range of Planck length of 1.6×10^{-35} m, where space-time is oscillating in an anarchic way, creating huge tidal effects. Thus, all matter entering the black hole is attracted toward its center and destroyed before reaching it by these enormous tidal effects.

Incidentally, the previous term of "Schwarzschild singularity" to refer to the sphere of Schwarzschild radius was misleading: it is not a singularity (no parameter being infinite), but an invisible border inside which things can only fall and nothing can escape; hence, it was replaced by the term "event horizon". Its circumference is equal to 18.5 km times the black hole mass expressed with the solar mass unit. Thus, the circumference of a black hole of ten solar masses is 185 km; the gravity field at the event horizon is 1.5×10^{11} times that on Earth! If the black hole is huge (>10,000 solar masses), a free-falling observer crossing the event horizon don't notice anything locally, which is consistent with the principle of equivalence: he sees everything behaving like previously, his heart beating at the same pace (proper time universality), photons near him go at the same speed c, and he still needs 3 minutes to boil an egg, etc. Conversely, observers outside the event horizon stop seeing him when he passes it: before then, they notice that his photons take longer and longer to be emitted, and incur more and more redshift. However, if the black hole is smaller than 10,000 solar masses, our observer feels a significant tidal effect when passing the horizon due to the important difference in gravity between his feet and his head: he is extremely strongly extended. Then when continuing toward the center, he will reach a speed very close to c, and feel unbearable extensions and compressions due to this tidal effect.

The next major problem was how to find black holes. Since they don't emit anything, scientists focused on the possible visible effects they could produce. First, there cannot be any black hole in the solar system, or else their important gravitation effects would have been noticed. The solution was found in 1966 by the Russian team headed by Y. Zeldovitch who had been working on the atomic bomb: they described two mechanisms which combined together could ultimately work:

- If a star is orbiting around a black hole, its velocity must be very high in order to offset the huge gravity field.
- If a black hole crosses a cloud of gas, when the particles meet again after circling the black hole, they must reach extremely high temperatures and trigger huge shock waves, generating very powerful radiation in the X-ray range.

This induced astronomers to look at important sources of X-rays which were close to stars, and thanks to the spectacular progress of X-ray telescopes

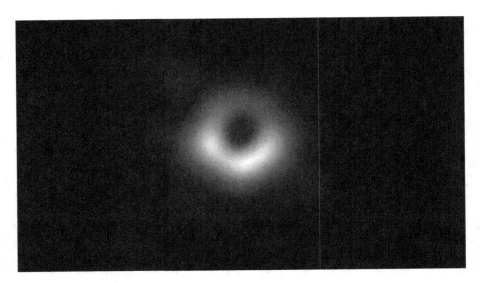

FIGURE 7.3
Black hole. (Image from Shutterstock. Credits: Event Horizon Telescope collaboration et al.)

in the sixties (due to their military applications), they found several black hole candidates. The first black hole, Cygnus X-1, was detected in 1964: it was the second-brightest object ever seen in X-rays, but it was not identified for certain as a genuine black hole until 1990. Cygnus X-1 is located at 6,000 light-years from the Earth and its mass is between 7 and 13 solar masses; its orbiting companion is a star between 20 and 30 solar masses (Figure 7.3).

Using the Event Horizon Telescope, scientists obtained an image of the black hole at the center of galaxy M87, outlined by emission from hot gas swirling around it under the influence of strong gravity near its event horizon.

Black holes are fully characterized by their mass, their rotating speed and their electrical charge. To date, many black holes of various kinds have been detected: many have several times the solar mass, some are spinning fast, some have an electrical charge, and others don't. In the 1990s, the center of our Milky Way was found to host a huge black hole, named Sag^4 A, with a mass of 3 million solar masses! Its horizon is 26 million km in diameter which represents 90 light-second and doesn't cause any problem for us since we are 27,000 light-years away. It later appeared that gigantic black holes probably exist in the middle of most big galaxies, some reaching 6.10^{10} solar masses!

Some of these supermassive black holes form quasars, which are huge rotating black holes surrounded by a gaseous accretion disk orbiting around them. As gas falls into the black hole, energy is released in the form of electromagnetic radiation. The power radiated by quasars is enormous, and they are the most luminous entities in the universe, some with a luminosity thousands of times greater than the Milky Way. Thus, ironically, the darkest objects generate the brightest ones.

Epilogue

Chandrasekhar and Fowler received the Nobel Prize in 1983.

Zwicky's predictions were correct: the first neutron star was observed in 1967. Oppenheimer was also correct: to date, the masses of the 3,000 neutron stars which have been observed range from 1.10 to 2.16 solar masses. Their diameters are between 20 and 40 km.

Zwicky was also correct as regards the mechanism for the creation of supernovae, and he was granted the Nobel Prize. Today we have a better understanding of the mechanism for their colossal ejection of matter: when a great star has consumed all its fuel, as Zwicky predicted, its core is turned into neutrons. This core is extremely hard and dense, and so when the particles in the upper layers are no longer sustained by the radiation and fall into it, reaching very high speeds (40,000 km/s), their crash generates a huge shock wave which blows away the external parts, scattering all sorts of matter and radiation into the cosmos. Thus, a star of initially 25 solar masses can blow away 24 solar masses while remaining a neutron star of 1 solar mass. In addition, another important idea has to be credited to Zwicky: he was the first to suggest the existence of dark matter.

The black hole, and in particular its singularity, generated intense studies and speculations which are still ongoing. Wheeler was a pioneer in quantum gravity, but the unity between quantum mechanics and General Relativity is still an important subject of studies. S. Hawking showed that black holes must emit a very small amount of radiation while reducing size extremely slowly and warming up, so that they will ultimately explode and disappear over an extremely long period. (A black hole of two solar masses will vanish in 10^{67} years!) Still many important questions remain open, some having interesting commonalities with the "Big Bang" as well as the possible future "Big Crunch" – both involve singularities with huge mass/energy concentrated in an extremely small region.

As for Landau, after spending 1 year in the Gulag, Stalin realized that he was precious for some strategic research. Hence, he survived and even earned the Nobel Prize. He supported Oppenheimer's opinion on the existence of black holes, which induced top Russian scientists, in particular Zeldovitch, to study them.

Schwarzschild unfortunately died in 1916, unaware of the immense consequences of his finding.

7.3 Gravitational Waves: First Observations One Century After Their Prediction

7.3.1 Theoretical Aspects

In 1916, Einstein predicted that an accelerating sizeable mass generates gravitational waves, just as accelerating electrical charges generate electromagnetic waves[5]. This is due to the General Relativity relation, which consists of differential equations of the second order involving time and space, like electromagnetism. Again like electromagnetism, gravitational waves don't need any matter (or ether) to propagate, and both propagate at light speed. They both have potentials which fade with distance by a ratio of 1/R. However, gravitational waves modify the very chrono-geometry of the space-time, whereas electromagnetic waves follow existing geodesic lines. Moreover, gravitational waves are not altered by the matter they cross.

Calculations show that, due to the extreme rigidity of our space-time universe, gravitational waves which are detectable on Earth must have been generated by extremely massive objects having extremely high accelerations, such as the coalescence of pairs of black holes or neutron stars (both being extremely dense and massive). However, the resulting space-time deformations at the level of our Earth are in the range of 10^{-18} meters, meaning a distance of one-ten-thousandth of the diameter of a proton!

Black Hole Coalescence: According to General Relativity, a pair of black holes orbiting around each other loses energy through the emission of gravitational waves, causing them to gradually approach each other for billions of years, while reducing their orbital period, and increasing their relative speeds. They reach extremely high speeds in the final minutes; then during the final fraction of second, the two black holes collide into each other at the speed of nearly 0.5 c, and form a single more massive black hole that spins at high speed. A portion of the combined black holes mass is converted into energy, and this energy emits a final strong burst of gravitational waves.

Figure 7.4 below shows the theoretical trajectories of a pair of black holes before their coalescence:

FIGURE 7.4
Coalescence of a pair of black holes. (©Swinburne Astronomy Productions/Swinburne University of Technology. Used with Permission.)

According to the wave-particle duality, gravitation is supported by the graviton particle, while the photon particle supports electromagnetism. The graviton has not been observed yet, but like the photon, its speed is assumed to be c and its mass zero, which is consistent with the very long range of gravitational effects. Gravitational waves are thus assumed to be due to gravitons emitted by large masses incurring huge accelerations.

Our Earth is orbiting around the Sun, and therefore also emits gravitational waves, but with an extremely small amount, so that its trajectory is virtually not impacted.

The existence of gravitational waves was first shown in 1974 by J. Taylor, R. Hulse and J. Weisberg who discovered a binary system composed of a pulsar orbiting around a neutron star. They observed that the orbital period of the pulsar was slowly shrinking over time, associated with two celestial bodies gradually approaching each other. This could be explained by a significant release of mass/energy in the form of gravitational waves.

The theory on gravitational waves made significant progress in 1982, showing in particular that the effects of the order of v^5 / c^5 account for gravitational waves. The theoretical decrease in the orbital period of this pulsar was of 67.10^{-9} seconds, which matched extremely well with what had been observed.

The effects of a gravitational wave are perpendicular to its propagation direction (when we are far from the source). As an illustration, a circle appears like an ellipse which successively enlarges and shrinks as the gravity wave passes, as shown in Figure 7.5.

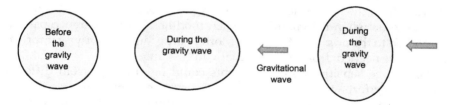

FIGURE 7.5
Effects due to a gravitational wave on a sphere.

Moreover, the gravitational waves generated by the coalescence of a pair of neutron stars or black holes could be modeled and then computed, which involved very complex equations and calculations. The form of the waves could thus be predicted, which was a major accomplishment.

7.3.2 First Gravitational Wave Detection on Earth in 2015

The first attempt to build a gravitational wave detector was due to the American engineer Joseph Weber in 1957. But the sensitivity requirements were so stringent that it was only in 2015 that the first gravitational wave detector, named LIGO, was able to directly detect gravitational waves on Earth, and this after huge investments (more than 1,000 scientists) as well as important technological and scientific breakthroughs.

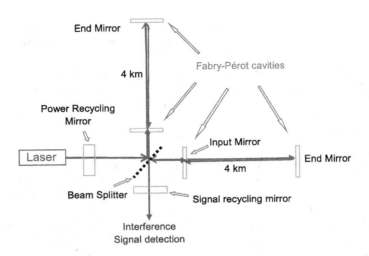

FIGURE 7.6
LIGO interferometer schematic.

The principle of the gravitational wave detector is based on the Michelson-Morley interference layout (cf. Figure 7.6): a gravity wave generates changes in the lengths of the interferometer arms, and these changes are different in the two arms since one arm is pointed closer to the direction of the gravity wave than the other one. Consequently, the phases of the two beams are not the same when recombining, which modifies the interference pattern. Compared to the original Michelson-Morley setup, huge sensitivity improvements were needed: from 10^{-7} m initially to 10^{-18} m!

These huge sensitivity improvements could be obtained from different sides: huge efforts were made to isolate the system from all sources of perturbations on Earth, to increase the signal resulting from the gravity waves and to reduce all sources of noise and fluctuations that are inherent in any instrument. In particular, the following was accomplished:

- In order to increase the signal, the interferometer arms are very long: 4 km. Moreover, Fabry-Perot cavities are used so that the beams travel a distance of 1,120 km by making several round trips between the mirrors before combining and interfering.

- In order to reduce all sources of noise and perturbations, the interferometer arms are built inside tunnels covered with a thick concrete framework. A vacuum is made inside the interferometer arms to avoid perturbations from the air.

- The mirrors are suspended with cables, and attached via special shock absorbers (seismic isolation).

- A complex system amplifies the power of the laser, using power recycling mirrors, and compensates for its small fluctuations in power as well as in frequency and geometry.

Last but not least, it was important to have two such interferometers separated by a large distance in order to make sure that the observed signal does not come from a local source of noise or perturbation, but from a common source: a gravity wave. LIGO thus comprises two distant wave detectors separated by 3,002 km, one in Hanford (Washington) and the other one in Livingston. Additionally, the time difference between the signals received in the two locations gives an indication as to the direction from which the wave came.

In September 2015, a few days after LIGO was brought into operation, a very clear and strong signal was captured, having all characteristics of gravitational waves. Both distant detectors detected the same signal with a delay of 7 ms, which corresponded to the propagation time of the gravitational waves between them. These waves were due to the coalescence of two black holes. The form of the signal received by LIGO gave a brilliant confirmation of the theoretical predictions (Figure 7.7).

This form is characterized by a frequency and an amplitude which simultaneously increase, both reaching a maximum at the coalescence of the two great masses; then both the frequency and the amplitude fade. Such frequencies are in the range of 30–200 Hz, and the signal lasts about 0.15 seconds. LIGO powerful algorithms took only 3 minutes to recognize the coalescence signal after it was detected.

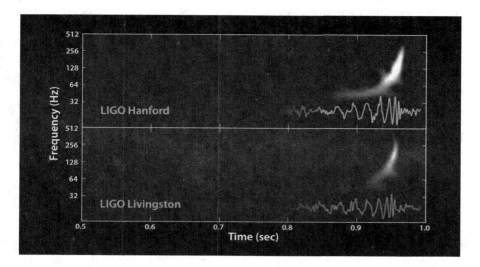

FIGURE 7.7
Signals received by both interferometers (LIGO/LSC).

LIGO scientists estimate that these waves were generated by an event which took place 1.3 billion years ago: the coalescence of a pair of black holes of about 29 and 36 times the mass of the Sun. In a fraction of a second, approximately 3 times the mass of the Sun was converted into gravitational waves – with a peak output power of about 50 times that of the whole visible universe. The phenomenon took place in the Southern Hemisphere, as indicated by the time difference in the wave arrival in Livingston and in Hanford.

7.3.3 A New Powerful Tool for Cosmology

Gravitational wave detection opens a new era of cosmos observation. Due to the significant number of galaxies (over 2,000 billion) and stars (our Galaxy contains 200 billion stars), it has been estimated that the yearly number of events which are comparable to this first one is between 200 and 400. And indeed, LIGO detected a second event 3 months later, similar to the first one, but without reaching the "5 sigma" statistical threshold required to be sure that it was not due to perturbations or noise. Besides mergers, gravitational wave detectors should be able to detect explosions of supernovae as well.

One first finding is the very existence of such massive black holes in the range of 30 times the solar mass, while those that were detected in our galaxy were much smaller, apart from the huge one in its center. It also was the first evidence of a merging pair of black holes. From 2015 to 2021, ninety such events were detected.

Compared with other observational means, gravitational waves give a more direct determination of the mass of the concerned objects. Furthermore, they enable us in theory to detect events that are not detectable by other means, like those that occurred before the recombination era. The Big Bang itself may have generated gravitational waves, as well as the huge inflation phase that occurred immediately after.

The importance of this new means of observation has received worldwide recognition: several even more powerful detectors are being constructed, in particular in Japan with the KAGRA project leading to the most sensitive detector. The European VIRGO project has been implemented in Italy; India is constructing a LIGO clone. These instruments will enable us to detect more events and at a greater distance, and to better qualify them. For the longer term, there are plans for launching a laser interferometer in space, which would orbit the Sun, enabling us to escape from our noisy environment on Earth, and to benefit from the possibility of having huge distances between the mirrors, such as 1 million km!

A parallel can be made with astronomers and scientists in the 17th century: they were the first in history to have precise telescopes enabling them to observe stars and planets. Their findings not only concerned the cosmos, but they also discovered in particular the law of inertia and Newton's laws, which are applicable to our day-to-day life. Thus, the new window opened by gravitational wave detectors to a totally unknown part of the universe may well also bring us some important knowledge not only on this very remote part of the cosmos, but also on our day-to-day reality.

Notes

1 In his memoirs, the émigré Russian scientist George Gamow reported that Einstein once described the cosmological constant as "my biggest blunder." From: https://blogs.scientificamerican.com/guest-blog/einsteins-greatest-blunder/.
2 Cf. Pb 4.3.
3 Cf. Volume II Problem 3.14.
4 Sag A is in the Sagittarius constellation.
5 Remark: The acceleration has to be dissymmetrical to generate waves.

8

Epilogue: A Scientific and Epistemological Revolution

Introduction

The long process through which the theory of Relativity was accepted is a sign that it was not only a scientific revolution, but also an epistemological one. The main steps that led to the theory of Relativity – including the major contributions of different scientists from Copernicus to Poincaré, who came very close to the result – are retraced in this epilogue. The dramatic changes in the scientific approach and methods that were initiated by the theory, and amplified during the 20th century, are then recalled. Ultimately, Albert Einstein's personality and history are analyzed as a potential key to understanding his contribution.

8.1 How Relativity Was Accepted

After publishing his two fundamental articles on Special Relativity in the prestigeous "Annalen der Physik", Einstein expected harsh criticisms, but the first feedback only came the following year, when Max Planck, one of the most renowned scientists worldwide, sent him a letter asking for some clarifications. Planck, together with a small group of scientists, then contributed to the development of this new theory. He even stated that the importance of the theory was "comparable in depth and in scope to the Copernican revolution[1]".

Nevertheless, it took a long time before the new theory was accepted, mainly because of the lack of experimental confirmations. The dominant opinion was expressed by the dean of professors at Bern university, when rejecting Einstein's application in 1907 for being assistant professor: "We believe that your conclusions on the nature of light and the connections between time and space are rather radical, and that your hypothesis come more from art than science[1] ».

Then in 1907–1908, two major events considerably boosted the acceptance of Relativity among the scientific community:

- A seminal article by Hermann Minkowski, who was a renowned mathematics professor at the prestigious Göttingen University. Minkowski famously said: "Space by itself, and time by itself, are doomed to fade away into mere shadows, and only a kind of union of the two will preserve an independent reality". He showed that we are in a 4D space-time universe, and introduced the fundamental concepts of proper time and events, as well as the new metric, all of which are extremely useful and meaningful even today.

- Another significant discovery by Einstein: the principle of equivalence, which is the cornerstone of General Relativity.

In 1909, Einstein's first academic recognition came when Geneva University granted him an honorary doctorate. The same year, he was nominated as a full-time "Extraordinary" Professor at the University of Zurich, and could leave his position that he had occupied rather grudgingly at the Swiss patent office.

In late 1913, the Prussian Academy of Sciences strived to attract Einstein. Planck's letter of recommendation for him reads as follows: "All in all, one can say that among the great problems, so abundant in modern physics, there is hardly any to which Einstein has not brought some outstanding contribution[2]".

General Relativity was the next important step: Special Relativity aimed at being universal, but gravitation seemed to be incompatible with the new theory. Gravitation was indeed thought to produce instantaneous, distant interactions between massive objects, even over huge distances, whereas Relativity implied that no speed, including distant interactions, could be faster than the speed of light. The key to solving this problem was the principle of equivalence, stating that gravitation is locally equivalent to an acceleration. Even though the mathematical translation of this general principle was extremely complex, Einstein succeeded in finding its solution in 1915. Later, during the solar eclipse of 1919, the results of a series of observations of the bending of light rays in the proximity of the Sun were shown to be in accordance with General Relativity. Einstein thus became famous worldwide, with a greater prestige than any scientist had ever enjoyed.

However, Einstein did not receive the Nobel Prize for his Relativity theory, but in 1921 for another important and revolutionary finding: the photoelectric effect. The Nobel Committee mentioned his theoretical research, but without quoting the word "Relativity". The Nobel Prize privileged experimental work over theoretical studies, and Relativity was viewed as belonging to that category and even as being rather philosophical; there was skepticism

---

[1] Quoted by Denis Le Bihan in his book, *L'erreur d'Einstein*, page 31.

among the jury as the Relativity theory was rather complex (very few people understood General Relativity) and not supported by much experimental evidence; no significant applications were foreseen at that time; last but not least, anti-Semitism also played a role: one member of the consultative committee, Philipp Lenard, publicly called Relativity "the great Jewish fraud[3]"!

When the Nazis came into power, they considered Relativity to be an insane product of "Jewish intellectualism", and prohibited teaching it. Some major German scientists, such as Heisenberg, however, wanted to keep using and teaching Relativity. As the Nazis did not want to admit that Relativity was discovered by a member of the "inferior race", mentioning the name of its discoverer became forbidden and Relativity was attributed to F. Hasenöhrl, an Aryan researcher. Incidentally, they also attributed Niels Bohr's work to an Aryan.

This long process for the recognition of the theory of Relativity, and the reaction from the Nobel Committee are clear signs that Relativity was not only scientific revolution, but also an epistemological one.

8.2 An Epistemological Revolution

During the two centuries before Relativity, science and technology had achieved so many impressive accomplishments that science was believed to be able to explain everything in the near future. Moreover, physical laws, such as the famous Newtonian ones which successfully explained the trajectories of planets as well as those of falling apples, were considered absolutely true. Science and truth were almost synonymous. Moreover, some famous scientists like Pierre-Simon Laplace believed that science would also be able to predict everything in the near future, since physical laws were deterministic. This approach, however, required knowing the initial conditions for our universe, which was problematic.

The previous scientific revolutions owed to Nicolaus Copernicus in the 16th century, and later Galileo and Newton, were considered to be victories of science over non-scientific approaches, mainly ruled by religion. Religion was very prevalent: major scientists, such as Galileo, were threatened with death when their scientific statements contradicted some religious dogma. It took more than 200 years for the Copernican revolution to be widely accepted.

Science became based on rational methods and rigorous logic; experiments and observations were "sacred". These were the reality from which scientists used their intelligence to induce new physical postulates and laws; these were also the reality with which their findings must be confronted. Yet, this ideal methodology suffered exceptions: Newton in particular was a strong believer, and his faith induced him to believe in the absolute frame (space and time) as being the theater where each phenomenon immediately resulted from the whole mighty power of God.

One century after Newton, another great scientist, Fresnel, postulated that light propagation was due to the vibrations of an invisible substance called the ether. He reasoned by analogy with sound, but no evidence of the existence of ether was provided. When these methodological issues came from such major scientists as Newton or Fresnel, the vast majority of the scientific community did not object, all the more so as these wrong assumptions reinforced each other. To quote French philosopher Michel de Montaigne, "Nothing is so firmly believed than that we know least[4]".

There were, however, a small number of scientists who expressed doubts and even harsh criticisms: notably Berkeley and Leibnitz from the outset, and then Ernst Mach at the time of Einstein. Mach was considered a heretic, but Einstein decided to follow him. Mach pointed out that Newton's and Fresnel's assumptions were not grounded on physical reality. His general view was that "all knowledge is a matter of sensation. What men delude themselves in calling 'Laws of nature' are merely summaries of experiences provided by their own—fallible—senses...these are the only realities. Others do not exist[5]". For Mach, simple concepts were created for economical reasons, to make it easier to describe complex realities. But as science progresses, these concepts should be challenged and evolve. Mach also influenced Einstein significantly as regards General Relativity with his view that inertial forces were due to the presence of distant great masses. Einstein recognized Mach as a main forerunner of Relativity, and praised "his incorruptible skepticism[6]". Ironically, Mach remained skeptical about Relativity.

Einstein was a revolutionary who appeared in a domain where no revolution was expected. One of the rare problems that were left open at the end of the 19th century concerned the disappointing results of the Michelson-Morley experiment: the ether had no effect on light propagation. Maxwell's equations would then lead to an absurdity: the speed of light would be independent of the speed of the source which had emitted it. Very few scientists thought that it would induce drastic changes; however, August Föppl was an exception and wrote in his book on Maxwell's theory: "A deep revision in our conception of space is perhaps the most important problem of science of our time[7]". Nobel laureate Hendrik Antoon Lorentz commented that Föppl deserved primary acknowledgments as a precursor of Relativity for his outstanding book that Einstein had surely read. Einstein carried out this revision in a very bold and in-depth manner, which led to the conclusion that the invariance of light speed was not absurd, but what was absurd was the set of postulates from which we deduced that it was absurd.

The scientist who came extremely close to the solution was the great French mathematician and scientist Henri Poincaré. He pursued an idea initially originated by George Francis Fitzgerald in 1889: the ether produces electromagnetic effects which reduce the lengths of objects. This assumption was not taken seriously by the scientific community until Lorentz pursued it and issued a formula expressing this length contraction. Poincaré completed this work and showed that time also was impacted (dilated) by the motion within the ether, and he thus expressed the new law of kinematics which became the corner-stone of Relativity, and which he elegantly called the "Lorentz transformation".

A few months later, Einstein stated the same law, which he had obtained by different means and with a different explanation. He rejected both the absolute frame and the existence of ether. Conversely, Poincaré kept assuming the existence of ether and Newton's absolute reference frame as giving what he called "true time[8]", and so did Lorentz. However, Poincaré distanced himself significantly from the absolute frame: he even prophetically expressed a few years before 1905 that "absolute space, absolute time, and even Euclidean geometry are not necessary conditions for mechanics[9]".

The mention of this new theory as the principle of Relativity is first owed to Poincaré. Einstein kept the name, but actually transformed the meaning of both "Relativity" and "principle".

- Regarding "Relativity", the author of this principle, Galileo, had in mind the Newtonian absolute frame; he stated that the laws of physics are the same in all inertial frames, but these frames were defined as being in inertial motion relative to the reference frame that hosted the fixed visible stars. Einstein was very bold to state that such a reference frame didn't exist, all the more as at that time, the known universe was reduced to the Milky Way.

- Regarding "Principle": a principle was considered as a general guideline that could eventually suffer exceptions. Conversely, Einstein considered the principle of Relativity as a postulate, meaning that no exception was possible. His position was based on his faith in a harmony in the universe, and he subsequently used this postulate in his demonstrations. The main scientists considered that he was not allowed to do so because this principle was to be demonstrated in every new context.

Einstein had cut the Gordian knot, resulting in a much simpler and clearer axiomatic framework. He was confident his theory was right despite the lack of experimental evidence, mainly because of its beauty in the sense that "the more impressive the greater the simplicity of its premises is, the more different kinds of things it relates, and the more extended is its area of applicability"[10] (Einstein). Simplicity, beauty and harmony are often marks of truth: the word "cosmos" has the same root as beauty (cosmetics). A brilliant illustration of this principle is given by Copernicus, as the main reason for his revolution was that it brought simplicity into the descriptions and calculations of the trajectories of celestial bodies. Such idea that the simplest explanation is usually the right one is actually very old: this principle was issued by William of Ockham in the early 14th century; it is known as "Ockham's razor" (that cuts all that is unnecessary) and is still in use today.

Contrary to appearances, it is not easy to make a system simple. When it has to take new needs into account, the natural way is to add complexity to that system.

After several iterations, it becomes a Gordian knot, which is not easy to destroy without proposing an alternative which is as efficient in the same domain of application. In our case, it was nothing less than physics in its entirety except for Maxwell's theory. Those who know the old system perfectly and have been working on it for a very long time are not inclined to destroy it, but to make it cope with the new requirements. This can explain Poincaré's approach, which was rather ontological, meaning trying to find explanations that are compatible with existing laws (postulates). Although Einstein was highly impressed by Poincaré, their relations were rare and somewhat fresh. Einstein said that prior to his 1905 publication, he didn't know Poincaré's paper that mentioned the Lorentz transformation. Conversely, Einstein always received good support from Lorentz, who wrote the following to him in 1915: "I felt the necessity of a more general theory, which I developed later in 1904, but which was actually formulated by you (and in a less extended manner by H. Poincaré)".

In 1908, Minkowski introduced major conceptual enhancements into the theory of Relativity, but Einstein initially considered these as "superfluous manifestations of mathematical erudition[11]". A few years later, however, he acknowledged that these were a key to his next revolution: General Relativity. Thus, Relativity was a revolution in several steps due to several revolutionary heroes: the groundwork was laid by Mach, Fitzgerald, Lorentz, Föppl and Poincaré; the decisive actions were taken by Einstein and, in a less disruptive manner, Minkowski. As regards General Relativity, once more the brighter ideas came from Einstein, who showed great perseverance when faced with vast mathematical complexity. He was helped by great mathematicians, in particular David Hilbert, Gregorio Ricci-Curbastro, Tullio Levi-Civita, and, last but not least, by his good friend M. Grossmann. Hilbert found the same General Relativity relation at approximately the same time as Einstein (even 1 week earlier), and by different mathematical means (Lagrangian). However, he did not claim ownership as he recognized that Einstein had explained all the physical aspects to him.

This revolution had important epistemological consequences: for the first time in history, a main postulate, light speed invariance, was going against the common sense. Intuition indeed plays a counterproductive role in Relativity, and Einstein explained that our common sense and our intuition stem from our experience, but no one has ever traveled at speeds close to light. For the first time in history, a new theory was neither grounded on observations nor on experiments. In the past, however, some physical laws had also gone against the common way of thinking, for instance the law of inertia, but not with the same disruptive intensity as Relativity.

As in many revolutions, Relativity was the theater of excesses and errors. In particular, it was often said in the early years of Relativity that "time doesn't exist", which irritated some philosophers, Bergson in particular. This sentence was far too excessive; Einstein did not pronounce it. Relativity restricted considerably the notion of time, and revealed its intimate link with space and even matter-energy. Stating that time did not exist was rather indicative of the growing importance of mathematics within society in general. Another wrong statement which is still commonly heard is that "all is relative". That is an oversimplification to impress the public. There are several other false statements that are still quite common and confuse not only the public but also many students, especially about the time dilatation.

More importantly, Relativity showed for the first time that science could be wrong. Science is the best representation of reality given by humankind at a given point of time. There will always be new observations, experiments, and even deeper understanding of existing laws which will lead to new theories and eventually invalidate previous ones.

Since Relativity was accepted, scientists have been encouraged to doubt postulates and laws. In addition, they have been elaborating new theories which are less based on experience and observations, but more on abstractions and, in particular, on mathematics. Einstein dramatically changed his views regarding this matter: he had initially said that neither Special nor General Relativity was speculative, and that both arose only after observation[12]. Conversely, later in his life, he said that the creating factor resides in mathematics, and that "only the theory decides what is observable". Heisenberg recognized that this fascinating guideline helped him find his famous uncertainty principle. However, mathematical predominance raises some issues: several fundamental physical laws were found thanks to mathematics before their physical explanation could be given. Schrödinger's equation is especially illustrative: Schrödinger and Einstein strongly rejected its explanation which soon became dominant, the Copenhagen interpretation. For the first time in science, quantum mechanics showed laws that went against scientific determinism. But, in this case, the revolution went further than what his initiator was prepared to accept: Einstein insisted that "God does not play dice with the universe[13]", to which Niels Bohr answered: "Einstein, stop telling God what to do[14]".

A major factor that reduces the roles of experiments and observations is that they have become much more difficult to achieve. Since Relativity, several important findings came from the theory and were much later confirmed by observations and experiments. The Big Bang theory originated from the General Relativity theory before any observations pointing to it. This was also the case of the Higgs boson, a key element in the Standard Model of particles, which was observed 48 years after it was postulated. Similarly, gravitational waves were observed 100 years after their prediction.

As science has progressed, it has also revealed the immensity of what we ignore. For example, we do not know 95% of the mass and energy that compose the universe (dark matter and energy). Science is not likely to explain everything in the near future. One important objective is to find a way to reconcile General Relativity and quantum mechanics despite the fact they are to be incompatible. In their search for new theories, scientists try to find the same simplicity and beauty as with Relativity.

As postulates become more and more speculative, one may wonder what actually distinguishes science from other disciplines (or even beliefs), since some postulates and some scientific results cannot be confronted with reality for practical reasons. Philosopher Karl Popper drew a red line: a scientific theory must be built on postulates and axioms which are refutable. This criterion is widely admitted among the scientific community. Paul Feyerabend, though, disagreed and pointed out that many important scientific findings were not based on scientific methods, for instance, the Copernican revolution.

As complexity has increased, scientists have become experts in very specific domains. At the time of Newton, Leibnitz and Descartes, scientists were polymaths, well versed in several disciplines (philosophy, mathematics, astronomy, religion, history, etc.). Science was part of the humanities, which favored a broader view, cross-fertilization and even a certain wisdom. Poincaré and Einstein were probably the last polymaths as they mastered a very large portion of their contemporary science.

Despite the growing importance of mathematics with its rigorous logic, various human and social factors have continued to play important roles in the work of scientists, as shown by the following examples:

- *Many major breakthroughs are due to very young men, even before they reach their thirties* (e.g., Einstein, De Broglie, Heisenberg, Pauli, Dirac, Chandrasekhar and Chadwick). Conversely, in the later part of their lives, scientists tend to be less creative, more conservative (like many people) and even to oppose new ideas. Einstein made no exception: he rejected the Big Bang theory, the existence of black holes and the Copenhagen interpretation of quantum mechanics.

- *The scientific community can still be wrong for a long time, black holes being an example.* : a young Indian science student, S. Chandrasekhar, showed in 1930 that Relativity makes it possible for black holes to exist in our universe. This was strongly and stubbornly rejected by the scientific community for more than 30 years. Why such opposition from the scientific community? Why couldn't they simply say: maybe, we don't know? A first explanation lies in the scientists' egos, and then in the courage that is necessary to oppose some leading scientists, such as Eddington or Einstein, after they take firm positions. Other reasons may stem from the fear of the unknown and the fear of death. Black holes are indeed terrifying: they rapidly annihilate everything that passes their invisible horizons. "Neither the sun nor death can be looked at with a steady eye[15]", as François de La Rochefoucauld wrote.

- *When confronted with the atomic bomb, the words of François Rabelais in the 16th century, "Science without conscience is nothing but ruin to the soul[16]", became more relevant than ever:* the way scientists behaved reflected the importance of both their personal opinions and the ideology of their regime: Otto Hahn, the head of the German team who first performed the uranium fission, stated that "God will not permit to make it[17]" (the atomic bomb). Many scientists, led by Leo Szilard, tried to reach an agreement within the international scientific community in order to keep that research secret, but it soon turned out to be impossible: some leading scientists published papers on fission hoping that their government would grant them more financing. Americans developed the bomb mainly because they were convinced that Germans were working on it, and that they would have no qualms about using it. These arguments strongly motivated American scientists. Conversely, Germans, who were certain that they were scientifically superior, were convinced that no other country was able to develop the bomb before the war came to an end. In the first part of the war, they were not receptive to an atomic project because it was based on non-Aryan science. Moreover, their harsh anti-Semitic policy deprived them of several top scientists (such as Lise Meitner, Enrico Fermi and Leo Szilard). Besides, they were confident that their military superiority would suffice. A couple of years later, they were eager to develop new devastating weapons, but several German top scientists were not Nazi fanatics, in particular Otto Hahn, as well as Heisenberg and Hilbert. Hence, they presented the bomb project as technically very complex, costly, long term and hypothetical, and they initiated a limited project of electrical power production. If Hahn had been a Nazi fanatic, the whole destiny of the world would have changed dramatically.

• Most American scientists were opposed to the use of the bomb against Japan without prior warning. After 1945, it became theoretically possible to make a super (thermonuclear) bomb, 1,000 times more powerful than Hiroshima's: most American scientists were reluctant to take part in this program until 1953 when they learned that the Soviets had taken the lead on it. There undoubtedly were many top-level scientists in the USSR, but initially, the communist party did not look favorably upon atomic science, which was considered not to be materialistic. Thus, like in Germany, ideology played a key role in the USSR, perverting science as exemplified by Trofim Lysenko. Conversely, immediately after Hiroshima, Russians decided that the atomic bomb was the absolute top priority. It is noteworthy that these major military programs greatly benefited such technology as computers and various high-precision instruments, which led to important scientific breakthroughs.

Thus, the gigantic step accomplished by Relativity actually knocked science off its pedestal: science is no longer about explaining everything; science is no longer synonymous with truth; the scientific community can still be wrong for a prolonged time; scientific methods are questionable; refutability has become uncertain; and scientists are now highly specialized. However, we can be sure that the "divine curiosity and playful drive of the tinkering and thoughtful researcher[18]" (Einstein) will continue to drive scientific progress.

8.3 Who Was Albert Einstein

Albert Einstein was born in 1879 to a German Jewish family who moved from Ulm (Wurttemberg) to Munich when he was 1 year old. His mother, Pauline Koch, was always very supportive; she loved literature and music. His father, Hermann, was a jovial, optimistic man, but not very successful in his small business of electrical devices. Einstein later said he was "exceedingly friendly, mild and wise[19]".

In his first years, Albert's brain had a peculiar way of functioning: it was only at the age of 9 that he could speak fluently and it often took him a long time before he answered questions. It has been suggested that he had a form of dyslexia. He had difficulties at school, and when his father asked his headmaster what career his son should pursue, the answer was: "It does not matter, he'll never make a success at anything[20]".

Albert was sent to the Catholic elementary school close to his home because Einstein's family was not Jewish observant. At that time in Munich most people were practicing Catholics; hence, he was confronted with several dogmas which he and his family considered fantasy and superstition. Albert then went to the gymnasium where he hated "the methods of fear, force and artificial authority. Such treatment destroys the healthy feelings, the integrity and the self-confidence of the pupils. All that it produces a servile helot[21]" (Einstein). But as Albert had a strong character, these methods had opposite results: he worked on his own and developed a radically inquisitive attitude.

Albert's passion for physics appeared at an early age. He was 12 when he realized from popular scientific books that most of the stories in the Bible couldn't be true. This was a big disappointment for him, and it produced two major consequences: again a rejection of all authority, and a search for something else to fill the vacuum, following in this respect the Jewish tradition "obsessed for centuries by a concept of order and harmony in the universal design[22]" (A. Eban). Einstein later wrote: "I want to know how God created this world... I want to know His thoughts, the rest are details[23]". Einstein also said: "I believe in Spinoza's God who reveals himself in the harmony of all that exists, and not in a God who cares about human activities[24]".

When Albert was 15, his family moved to Milan after his father went bankrupt and was rescued by his family-in-law. Albert, however, was left alone in Munich to finish his studies at the gymnasium, but he was soon expelled with the following comment: "Your presence in the class is disruptive and affects the other students[25]". It is interesting to note that he did not suffer from anti-Semitism at that time. Albert subsequently moved to Milan, and was very happy to find the people to be highly civilized and educated; he also appreciated Italy's art and freedom. Furthermore, he rejected his German citizenship, preferring to be stateless. Yet, his parents were desperate as he had no diploma, thus no possibility to enter university. Fortunately, there was an engineering school, and even the best one in central Europe except for Germany, which did not require any diploma, but selected its students through a difficult entrance examination: the Zurich Polytechnic. Egged on by his parents, Albert tried the examination at the age of 16, that is, 2 years in advance, and got an exceptional high grade in math, but very low ones in other subjects, which led the Director of Polytechnic to give him the chance to enter the following year after studying in a Swiss school in order to catch up in the subjects in which he had underperformed. So Albert eventually entered the Zurich Polytechnic and very much appreciated its friendly environment. He loved Switzerland and several years later became a Swiss citizen. Much later in his life, he said Switzerland was the country where people are "by and large more humane than the other people among whom I had lived[26]".

However, Albert was not entirely at ease at the Polytechnic: the courses were oriented toward engineering, whereas he was passionate about theoretical physics. As an example, Maxwell's theory of electromagnetism was not taught. But Albert once again worked on his own and later said it was "the most fascinating subject". Besides, he often skipped the courses which didn't interest him, all of which resulted in the fact that his teachers did not like him, even though he was awarded a diploma. His main professor said: "he is someone intelligent but to whom one can't say anything[27]". Minkowski, his professor of mathematics, who later became famous for his major contributions to the theory of Relativity, remembered Einstein as "a lazy dog, who never bothered about mathematics at all[28]". Consequently after the Polytechnic, Einstein did not succeed in embracing the academic career he had contemplated. He even had some difficulties finding a position in the Swiss Patent Office as a preliminary examiner of patent applications. Nevertheless, he was pleased to have enough spare time for his scientific research, which he carried out together with a small group of friends, in particular Michele Besso who shared his passion and remained his friend for the rest of his life. In addition, he later said that his position in the patent office had some positive aspects: he learned how to express himself, and was not forced to publish many articles on minor subjects as academic researchers often do.

It thus appears that from an early age, Einstein was prompted to challenge dominant beliefs and authority; hence, when addressing the issues of electromagnetism, it is understandable that he chose to listen to Mach's skeptical, but reasonable arguments, although the latter was considered to be a heretic.

In 1905, Einstein published five articles, two on Relativity, a revolutionary one on the photoelectric effect, building the foundation of quantum mechanics, and another very important one on the Brownian movement which gave J. Perrin the possibility to prove in 1928 the very existence of molecules and atoms. Hence, 1905 was called his "annus mirabilis". Nobel laureate Max Born wrote: "Even if Einstein had written nothing on Relativity, his other contributions to theoretical physics are so fundamental that they would be enough to make Einstein one of the greatest physicists of the history of thought[29]".

As a professor, Einstein was rather informal, close to his students, doing his best to make them understand what he taught. He even used to prolong his classes with some students in a café, writing on tablecloths, and so on.

Later, despite his detestation of Germany, he accepted an offer from the Prussian Academy of Sciences in Berlin, mainly because he held Planck in high esteem. Moreover, the offer came at a time when he was overwhelmed by very complex research on General Relativity, and the position made it possible for him to focus entirely on his research while providing him with high-level scientific support. Finding the equations of General Relativity took Einstein 8 years of very hard work interspersed with periods of errors and discouragement, while competing at the end with David Hilbert, the best mathematician of the time, who was also striving hard to find these equations.

His beliefs were a strong driver in his work: "True scientific thought is not possible without faith in the inner harmony of our universe, and from this axiom I developed my theory of relativity[30]". However, these beliefs hampered his readiness to accept the Big Bang, black holes and quantum mechanics: after initiating a revolution in quantum mechanics with his prophetic statement that the photon is the quantum of light, Einstein was reluctant to admit the probabilistic nature of fundamental physical laws. He was, however, an early supporter of the original idea in 1924 of Louis de Broglie, a French student, which soon became a pillar of quantum mechanics, that is, that particles also have wave aspects. Einstein then progressively moved away from quantum mechanics and tried unsuccessfully to bring together General Relativity and electromagnetism.

To the question of where his genius came from, Einstein answered: "I have no particular talent, I am merely extremely inquisitive[31]". In addition, some of his American colleagues were impressed by his strong sense of perseverance which they attributed to his German background.

Some other personality traits are also worth mentioning: Einstein was always a very determined pacifist: "Warfare cannot be humanized, it can only be abolished[32]". When he was in Berlin during World War I, he took very courageous pacifist positions, unlike most scientists who became active in the development of new weapons. He was sickened by the huge number of casualties and pleaded in favor of conscientious objection. In 1928, he became the president of the Human Rights League.

Despite his Nobel Prize, Albert Einstein lived under the threat of death, and was forced to flee from his native country as early as 1933. The Nazis had two reasons to hate him: he was born a Jew and was always an active pacifist. The USA welcomed him, and he became a professor and researcher at Princeton University. In 1939, Einstein learned that a secret uranium project was underway in Germany and subsequently signed a letter to President Roosevelt warning him that an incredibly devastating atomic bomb could be made, and that the Germans might get it. Einstein was not asked to take part in the development of the atomic bomb due to his strong pacifist convictions. At the end of the war, Einstein wrote another letter to the President of the

USA asking him not to use the bomb against Japan, and he deeply regretted his first letter. After the war, Einstein campaigned for worldwide nuclear disarmament, and famously said: "I do not know with what weapon World War III will be fought, but World War IV will be fought with sticks and stones[33]".

Einstein was also a Zionist who supported the creation of a state for the Jewish people in their historical land. He was offered the presidency of Israel, an honorific function according to the constitution of that State, but he declined it, explaining that he had "been dealing with the world of objects and that he had neither the ability nor the experience necessary to deal with human beings and to carry official functions[34]". Yet, he remained attached to the Jewish tradition: "The pursuit of knowledge for its own sake, an almost fanatic love of justice, and the desire for personal independence, these are the features of the Jewish tradition which make me thank my stars that I belong to it[35]".

Einstein was a very good violinist and loved music. His elder son said: "Whenever he felt that he had come to the end of the road or into a difficult situation in his work, he would take refuge in music, and that would usually resolve all his difficulties[36]".

Einstein rejected the social codes regarding his appearance, his dressing, his long hair, etc. Once, he paid a visit to a teacher of his old gymnasium in Munich, of whom he had fond memories, but the teacher didn't open the door to him when he saw someone dressed up like a beggar or a thief.

Einstein had strong needs for independence. For instance, he surprisingly said in 1954: "If I were a young man again and had to decide how to make a living, I would not try to become a scientist or a scholar or a teacher. I would rather choose to be a plumber in the hope of finding that modest degree of independence still available under present circumstances[37]". However, his passion for theoretical physics never abandoned him until his last day.

Notes

1 Quoted in M. Planck's lecture at Columbia University in New York, April–May 1909.

2 Quoted by Ronald W. Clarck in his book: *Einstein, the Life and Times*, page 169.

3 Cf. https://christiansfortruth.com/1905-nobel-prize-winning-german-physicist-philipp-lenard-science-is-racial-and-conditioned-by-blood/. His anti-Semitism and German nationalism were at the core of his opposition to Einstein's Theory of Relativity, which he branded as "the great Jewish fraud". A similar version is in the book by Clarck, *Einstein, The life and Times*, page 169.

4 Written by F. de Montaigne in his book: *The Complete Essays*. Cf. www.goodreads.com/quotes/198179-nothing-is-so-firmly-believed-as-that-which-we-least.

5 Cf. *Einstein, The Life and Times* by R.W. Clarck, page 38.

6 Cf. *Einstein, The Life and Times* by R.W. Clarck, page 38.

7 Cf. *Einstein, The Life and Times* by R.W. Clarck, page 154.

8 Cf. *Once upon Einstein*, by T. Damour, page 44 (French edition).

9 Quoted by Ronald W. Clarck in his book: *Einstein, The life and Times*, page 37.

10 From a quotation by Ronald W. Clarck in his book: *Einstein, The life and Times*, page 109.

11 Cf. Ronald W. Clarck in his book: *Einstein, The life and Times*, page 122.

12 Cf. Ronald W. Clarck, *Einstein, The life and Times*, page 154: Special Relativity was the outcome not of metaphysical speculation, but of considering scientifically the results of experimental evidence; also in page 33: the need for it (the theory) arose only after observation.

13 The Born-Einstein Letters 1916–1955. The quote made its first appearance in the Fifth Solvay International Conference.

14 Einstein liked inventing phrases such as "God does not play dice" and "The Lord is subtle but not malicious". On one occasion Bohr answered, "Einstein, stop telling God what to do". Cf. https://history.aip.org/exhibits/einstein/ae63.htm#:~:text=Einstein%20and%20Bohr,telling%20God%20what%20to%20do.%22

15 Written by the French writer François de La Rochefoucauld, quotation #8224. https://www.dicocitations.com/citations/citation-8224.php.

16 Written by the French writer François Rabelais in his book *Pantagruel*, II, 8.

17 Quoted in the book "*Heller als tasend sonnen*" by Robert Junck, French translation, page 71.

18 Quoted in the *Ultimate Quotable Einstein*, by Alice Calaprice, page 381.

19 Quoted by Ronald W. Clarck in his book: *Einstein, The Life and Times*, page 42–43.

20 Quoted by Ronald W. Clarck in his book: *Einstein, The Life and Times*, page 10.

21 Quoted by Ronald W. Clarck in his book: *Einstein, The Life and Times*, page 13.

22 Quoted by Ronald W. Clarck in his book: *Einstein, The Life and Times*, page 17.

23 Quoted by George Braziller in his book: *Einstein's 1912 Manuscript on the Special Theory of* Relativity, page 27.

24 Albert Einstein to Rabbi Herbert Goldstein (1929).

25 Quoted by Ronald W. Clarck in his book: *Einstein, The Life and Times*, page 20.

26 Quoted by Ronald W. Clarck in his book: *Einstein, The Life and Times*, page 42.

27 Quoted by Ronald W. Clarck in his book: *Einstein, The Life and Times*, page 39.

28 Quoted by Ronald W. Clarck in his book: *Einstein, The Life and Times*, page 120.

29 Cf. *Once upon Einstein*, by T. Damour, page 65 (French edition).

30 Quoted in the book *Einstein's 1912 Manuscript on the Special Theory of Relativity* by George Braziller, page 27.

31 Einstein to Carl Seelig, March 11, 1952, AEA 39-013, New York: Random House, 292.

32 Cf. *Einstein's 1912 Manuscript on the Special Theory of Relativity* by George Braziller (page 27).

33 Quoted in the book *Einstein's 1912 Manuscript on the Special Theory* of Relativity, by George Braziller, page 27.

34 Quoted in the book *Einstein's 1912 Manuscript on the Special Theory* of Relativity, by George Braziller, page 27.

35 Quoted in the book *Einstein's 1912 Manuscript on the Special Theory* of Relativity, by George Braziller, page 27.

36 Quoted by Ronald W. Clarck in his book: *Einstein, The Life and Times*, page 106.

37 Written by Albert Einstein on November 1954, in a letter to a magazine. Cf. https://www.localrooterandplumbing.net/post/albert-einstein-said-if-he-could-do-it-all-again-he-would-be-a-plumber.

9

Solutions to Questions and Problems

DOI: 10.1201/9781003201335-9

9.1 Answers to Questions and Problems of Chapter 1

Problem 1.1: Train scenario

Q1: There is no way for you to know if your train is moving because your train is an inertial frame, and the Earth too (as an approximation), and so all experiments made in your train or on the Earth will give the same results, according to the inertial frames equivalence principle.

Q2: 300 km/h, according to the inertial frames equivalence principle (cf. more in Volume II Section 1.2.2.3).

Q3: No, because 600 km/h is the result given by the wrong classical speeds additive law, being based on the absolute frame. However, the difference with the correct law is extremely small for speeds like 300 km/h (as we will further see).

Q4A: No, the beam is perpendicular to you in your train frame; it is then in an oblique direction in the Earth frame.

Q4B: Before Galileo, the Earth frame was privileged, and the motion of the train would have had no effect on the beam because after leaving the train, there was no force that would have dragged the beam in the direction of the train. Hence the beam would have been seen perpendicularly to the ground.

Problem 1.2: Train scenario: there is a tunnel 3,000 m ahead of the train

Q1: No, because of the relativity of distances.

Q2: No, because I am moving relative to the Earth where this segment of rails is fixed and measures 3,000 m.

Q3: Yes it is, because my friend is fixed relative to this segment of rails.

Problem 1.3: Train scenario: you and your friend emit light pulses toward the tunnel

Q1: No, he will find that your beam goes at the same speed as his, in accordance with the light speed constancy postulate.

Q2: The two laser beams will simultaneously hit the tunnel because your friend sees them going at the same speed and covering the same distance (cf. more on the event coincidence invariance in Volume II Section 1.3.1.1).

Q3: No, my beam will take a different time due to the time relativity.

Q4: Before Einstein, Q1 = Yes. Q2 = No. Q3 = No.

Problem 1.4: Train scenario: how much time will it take you to reach the tunnel?

Q1: 3/300 = 0.01 hour = 36 seconds.

Q2: No, because of the relativity of time between two events.

Q3: The duration (t2 – t1) is a proper time because these two times are taken by a person looking at his own watch and co-located with the events E and F. Conversely, your friend is not co-located with the event F; hence, his time measurement is not a proper time.

Problem 1.5: Train scenario: time inside the train versus time on the ground

Q1: No, you should ask for 3 minutes, exactly like on Earth, because the inertial frames equivalence principle states that all physical experiments give the same result in all inertial frames (cf. more in Volume II Section 1.2.2.4). There is no reason to consider boiling eggs as not a physical experiment.

Q2: The 3 minutes is a proper time for you in the restaurant car.

Q3: For your friend, the 3-minute eggs boiling time in the moving restaurant will not appear to be 3 minutes, due to the time relativity.

Problem 1.6: Universal homogeneity

Q1: A spherical universe cannot be homogeneous nor isotropic because its center is a privileged point, and also the points that are on its border.

Q2: The isotropy is violated because each point has a privileged direction: the direction toward the center.

9.2 Answers to Questions and Problems of Chapter 2

Problem 2.1: Appearance of a cubic station from a fast space ship

Shape b because the segments of the station that are parallel to the velocity v are seen contracted, while the transverse ones are seen unchanged.

Problem 2.2: Some useful relations:

$$\gamma^2\left(1-\beta^2\right)=\frac{1-\beta^2}{1-\beta^2}=1; \text{ then: } 1-\beta^2=\frac{1}{\gamma^2}.$$

$$\sqrt{\gamma^2-1}=\sqrt{\frac{1}{1-\beta^2}-\frac{1-\beta^2}{1-\beta^2}}\ =\ \sqrt{\frac{\beta^2}{1-\beta^2}}=\beta.\gamma.$$

Det $(\Lambda) = \gamma\,\gamma - (-\gamma v)(-\gamma v/c^2) = \gamma^2(1-\beta^2) = 1.$

Problem 2.3: Surprising effects when traveling in the cosmos

Q1: The 10 years duration is measured with Bob's proper time; hence, it is seen dilated by Alice who is moving relative to Bob at the speed 0.943 c, representing the γ factor of 3. Consequently, Alice's clock will show: $3 \times 10 = 30$ years.

Q2: Let's call E the event "Bob leaves Alice", and F the event "Bob arrives at Proxima". The time difference between these two events is a proper time for Bob who is present in both events, but this is not the case for Alice, meaning that it is not a proper time for Alice.

Q3: The 30 years duration of Alice is measured with her proper time; hence, it is seen dilated by Bob by the γ factor of 3; hence, Bob's clock will show $3 \times 30 = 90$ years!

Remark: It will take even longer for the call to reach Bob due to the transmission delay (the rocket's speed being close to c). Thus, there are serious limitations in the communication possibilities between Earth and extremely fast travelers!

Q4: The two events "Alice initiates a call at her year 30" and "Bob arrives at Proxima" are simultaneous in Alice's frame, but not in that of Bob who already left Proxima when Alice initiates her call.

Q5: Seen from Bob: Bob has traveled for 10 years at the speed 0.943; hence, he can state that he has covered a distance of 10 years × 0.943 = 9.43 light-years.

Seen from Alice: Bob has traveled for 30 years at the speed 0.943; hence, she can state that he has covered a distance of 30 years × 0.943 = 28.29 light-years.

The length Earth-Proxima measured by Alice is the proper length Earth-Proxima because she is fixed with the segment Earth-Proxima. We can see that our results are consistent with the length contraction law, which states that the proper length (28.29 light-years) is seen contracted by Bob (9.43 light-years) who is moving relative to this segment.

Problem 2.4: Absurdity of writing $t' = \gamma_v t$ **in the general case**

Let's apply the relation $t = \gamma_{-v}t' = \gamma_v t'$ to the times of the events E* and F*. We have in K: $t_F = \gamma_v t'_F$ and $t_E = \gamma_v t'_E$, meaning that $t_E = t_F$, which contradicts the relativity of simultaneity (cf. Einstein's train scenario where two simultaneous events in the train, K', are not seen simultaneous on the ground, K).

Problem 2.5: The image in K' of a straight line in K is a straight line

The origin events of K and K' being arbitrary, let's choose a common origin event O* which is seen in K at a point O which is on that straight line. Consider in K a 4D point A which is on that straight line. For any 4D point M along that straight line, we have: $\overrightarrow{OM} = k. \overrightarrow{OA}$. In K', the event A* is seen in A' and the point M in M'. Consequently, we have: $\overrightarrow{O'M'} = \Phi(\overrightarrow{OM}) = \Phi(k.\overrightarrow{OA}) = \Phi(k.\overrightarrow{O'A'})$, which shows that the points M' also form a straight line in K'.

Remark: This result is consistent with the law of inertia, which states that any object (including the photon) which is submitted to no force is moving along a straight line and at constant velocity in any inertial frame.

Problem 2.6: More on the Rossi & Hall Experiment at Mount Washington

Q1: Seen from a muon frame, the Earth is seen moving at the speed of 0.992 c toward itself.

Q2: The distance to the Earth of a muon at the altitude of 1,910 m is seen contracted in the muon's frame due to the length contraction law. Hence, the distance to the crash is: $(1,910 - 3)/\Upsilon$. Therefore, the travel duration is: $\tau = (1,910 - 3)/ (\gamma.0.992c) = 0.809$ μs.

Q3: This duration is measured in the same frame as the muon; hence, it can be directly compared with the muon disintegration period. Consequently, the same reasoning as in Section 2.3.1.1 applies: this duration represents $0.809/1.53 = 0.529$ times the disintegration period, leading to the theoretical number of muons at the low altitude of 3 m: $563. (1/2)^{0.563} = 390$ muons.

Problem 2.7: Check that the Lorentz transformation implements the time dilatation law

Q1: The time lapse (t2 − t1) is a proper time since the clock is fixed in K, which is not the case of K'.

Q2: In K, the events E* and F*, respectively, have for coordinates E (a, t1) and F (a, t2). Thus, the space-time separation vector E*F* has for coordinates in K: EF = (a–a; t2–t1) = (0; t2 − t1).

In K', the events E* and F*, respectively, have for coordinates E' (m', t'1) and F' (n', t'2). Subsequently, the space-time separation vector E*F* has for coordinates in K: E'F' = (n' − m'; t'2 − t'1).

Q3: We have: $(EF)' = (\Lambda).(EF)$, which means:

$$\begin{pmatrix} m'-n' \\ t'2-t'1 \end{pmatrix} = \gamma \cdot \begin{pmatrix} 1 & -v \\ -v/c^2 & 1 \end{pmatrix} \begin{pmatrix} O \\ t2-t1 \end{pmatrix}$$

Q4: The second line yields: $t'2 - t'1 = \gamma.(t2 - t1)$. It expresses the fact a in K, (t2 – t1), is seen dilated in K′ by γ, which is the time dilatation law.

Problem 2.8: Check that the chronological order between two co-located events is invariant

The previous exercise concluded that: $(t'2 - t'1) = \gamma.(t2 - t1)$ (1). In our context, (t2 – t1) is positive because the event E* is seen prior to F* in the frame K where both events are co-located. Consequently, relation (1) shows that the sign of $(t'2 - t'1)$ is also positive, meaning that E* is also seen before F* in K′. Thus, the absurd case where F* would be seen prior to E* in another inertial frame K′ cannot happen.

Problem 2.9: The Lorentz transformation implements the length contraction law

Q1: L is the proper length of the ruler because the ruler is fixed in K.

Q2: The observer in K′ needs to assess the abscissa of A′ and B′ at the same time in K′; otherwise, B′ would have covered more distance than A′; hence, (b′ – a′) would be meaningless (cf. Section 1.4.2.2).

Q3: $A' = (a', t')$ and $B' = (b', t')$; then: $AB = (b' - a', 0) = (L', 0)$.

Q4: No. **Q5:** Then: $A = (a, t1)$ and $B = (b, t2)$; hence,
$AB = (b - a, t2 - t1) = (L, t2 - t1)$.

Q6: $(A'B') = (\Lambda)(AB)$, which means:

$$\begin{pmatrix} L' \\ 0 \end{pmatrix} = \gamma \cdot \begin{pmatrix} 1 & -v \\ -v/c^2 & 1 \end{pmatrix} \begin{pmatrix} L \\ t2-t1 \end{pmatrix}$$

Q7: Using the reverse Lorentz matrix, Λ^{-1}, we have:

$$(AB) = (\Lambda^{-1}).(A'B') .$$

The reverse Lorentz matrix is obtained by the same matrix, but with the opposite velocity: -v instead of v. This means:

$$\begin{pmatrix} L \\ t2-t1 \end{pmatrix} = \gamma \cdot \begin{pmatrix} 1 & v \\ v/c^2 & 1 \end{pmatrix} \begin{pmatrix} L' \\ 0 \end{pmatrix}.$$

Calculating the first line (the space part), we obtain: $L = \gamma . L'$; then: $L' = L/\gamma$, which expresses the length contraction law. The ruler is indeed seen smaller in the frame K' than in K where it is fixed.

Problem 2.10: Axiomatic questions on our demonstration of the Lorentz transformation

The time dilatation law requires the light speed invariance postulate. Consequently, our demonstration of the Lorentz transformation requires the light speed invariance postulate.

It is not a problem that our demonstration of the Lorentz transformation does not require the length contraction law because the latter is a consequence of the same postulates used for demonstrating the Lorentz transformation.

Problem 2.11: Einstein's train scenario: typical time difference between the two pulses seen from the train

In K, the coordinates of E* and F* are: E(0, 0) and F(Xb, 0). The value of Xb is equal to the length of the segment AB in K, which is equal to the train length at rest (i.e., its proper length). We then have: EF = (AB, 0).

In K', the coordinates of E* and F* are E'(0, 0) and F'(X'b, T'b). The value of X'b is equal to the length of the segment AB seen in K'. We then have: E'F' = (A'B', T'b).

The Lorentz transformation from K to K' applied to E*F* gives for the time part: $T'b = -\gamma.AB.v/c^2$ (1).

With the values of the problem statement: The proper length of the train is 200 m, then AB = 200; v = 300 km/h. Relation (1) says that the passengers inside the train consider that the pulse B was emitted before the one in A, and that the time difference between them is: $T'b \approx 200 \times 300 \times 1000 / 3600.c^2 \approx 1.8 \cdot 10^{-13}$ seconds, which is extremely small, but still detectable with high-precision clocks.

Problem 2.12: Minkowski diagram with one frame: the Doppler effect

Q1: No, since the star is moving relative to the Earth, and Te is measured in K.

Q2: This is the same diagram as in Section 2.4.1, and it shows that $Te = To/(1 + \beta)$.

Q3: The proper period τ_e is seen dilated on Earth; hence, $\tau_e = Te/\gamma$;

then $\tau_e = \dfrac{To}{\gamma.(1+\beta)}$, *which matches the relativistic Doppler formula.*

Problem 2.13: Minkowski diagram with 2 frames: Einstein's train scenario

Q1: No, OB is not a proper length because the train is moving relative to the ground.

Q2A: See in the diagram below. **Q2B)** The flash emitted from O follows the line of light. The one emitted from B follows a perpendicular direction since its equation in K is $x = Xb - ct$ (with Xb being the abscissa of B* in K, meaning OB*) (Figure 9.1).

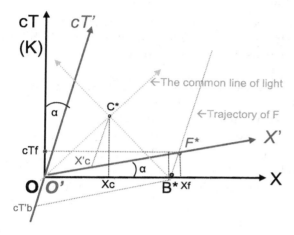

FIGURE 9.1
Minkowski diagram of Einstein's train scenario.

Q3A: See in the above diagram. **Q3B)** $Xc = Xb/2$ due to the symmetry of the pulses trajectories.

Q4: The trajectory of F is parallel to OcT' since F is seen moving in K at the speed v, like the point O'. Besides, the trajectory of F passes by the point B* since the point F was seen in B* at $t = 0$ in K.

Q5A: O'F* is the proper length of the train since O' and F* are the two extremities considered at the same time $t' = O$ in the frame K' where the train is at rest.

Q5B: No, because both extremities O* and F* are not simultaneous in K.

Q5C: It is OB* from the scenario statement: O and B are indeed seen simultaneously in K.

Q6: The event C* occurs closer to the back of the train than to its front. This is in accordance with what observers on the ground say, and since they did not make any implicit assumption, they are correct. Besides, the linearity of the transformation ensures that if one distance is smaller than the other one in one frame, it is smaller in all frames (assuming these distances are along the same direction).

Q7: The time at which this flash occurs in K is given by the ordinate of F in K, which is cTf in the diagram (with the segment F-cTf being parallel to OX).

Q8: Before, as seen by the point cT'b in the diagram, which is in the intersection of O'cT' and the parallel to O'X' passing by B*.

Problem 2.14: Time dilatation and de-synchronization effect

Q1A: The time t of the ground K is given from the one in the train by the reverse Lorentz transformation; thus: $t = \Upsilon.(t' + v/c^2\, x')$. In K', the two consecutive clocks show the same time: $t' = 0$; hence, the clock B is seen at the time $Tb = \Upsilon.\, v/c^2\, AB$.

Q1B: Tb is positive since $x' = AB$ has the same sign as v. Besides, in K the clocks A and B are seen beating at the same pace; hence, the clock B will always display a time which is later than the time of A.

Q1C: We can make the approximation $\Upsilon = 1$; 300 km/h ≈ 83.3 m/s; then: $Tb = 83.3 \times 3/9 \times 10^{16} = 2.8 \times 10^{-15}$ s. This explains why we did not notice the relativity of time.

Q2: The time difference is the same since the relation $t = \Upsilon.v/c^2\, x'$ is a linear function of x'.

Q3: See the diagram below and the comments (Figure 9.2).

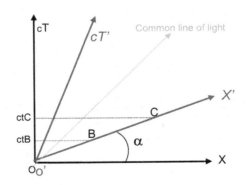

FIGURE 9.2
Minkowski diagram showing the time de-synchronization.

The clock A is at the point O = O'; the clock B is on the O'X' axis at the time $t' = 0$ in the train, and the same for the clock; besides, the clock B is in the middle of O'C.

From the ground (K), the clock B is seen displaying the time Tb, with cTB being the ordinate of B in K. The same applies for the clock C. Geometrically, cTb is in the middle of O-cTC.

Q4: From $Tb = \gamma.v/c^2\ AB$, we have: $\gamma.\beta = c.Tb/AB = 3 \times 10^8 \times 1/3 = 10^8$. We can see that β is very close to 1; hence: $\beta = 1 - \varepsilon$. Then: $\gamma = [1-(1-\varepsilon)^2]^{-1/2} \approx (2\varepsilon)^{-1/2}$. Then,

$$\varepsilon \approx \frac{1}{2\gamma^2} \approx 0.5. \times 10^{-16}.\ \text{Then}\ \beta = 1 - 5 \times 10^{-17}, \text{so finally: } v \approx c \times (1 - 5 \times 10^{-17}).$$

Of course, no train can go at this speed, but we can see that even with very fast rockets, we won't experience such an effect.

Problem 2.15: The Doppler effect

We have: $\lambda_0 = c\ \Delta\tau_0$ and $\Delta\tau_e = \Delta\tau_0. \dfrac{\sqrt{1-\beta}}{\sqrt{1+\beta}}$; then $\lambda_e = \lambda_0 \dfrac{\sqrt{0.15}}{\sqrt{1.85}}$, so:

$\lambda_e = 3400 \times 0.285 \times 10^{-10} = 968 \times 10^{-10}$ m.

The frequency received is $1/\Delta\tau_0 = c/\lambda_0 = 8.82 \times 10^{14}$ Hz.

The frequency emitted is: 3.10×10^{15} Hz.

Problem 2.16: Length contraction demonstration presented as an exercise

Q1: The observer on the platform considers that during (t2 − t1), the train has covered a distance which is equal to its length, noted L; he then deduces that the train length is: $L = (t2 - t1).v$ (1).

Q2: Observers inside the train deduce that their train has covered during (t′2 − t′1) a length which is equal to the length of their train; thus: $L' = (t'2 - t'1).v$ (2).

Q3: L' is the proper length of the train, denoted by L_p (the passengers being fixed in the train frame).

Q4: The two events E and F occur at the same place in the frame K, at the point A, so that (t2 − t1) is a duration of the proper time in A.

Q5: Then, according to the time dilatation law, the proper time duration in A is seen dilated by the passengers of the moving train relative to A. Thus: $t'2 - t'1 = \gamma_v.(t2 - t1)$ (3).

Q6: From (1), (2) and (3): $L' = L_p = (t'2 - t'1).v = \gamma_v.(t2 - t1).v = \gamma_v.L$. So finally: $L = L_p/\gamma_v$, which expresses the length contraction law.

Problem 2.17: The simultaneous light pulses

Q1: Both pulses move along the segment OB* on the common line of light (Figure 9.3).

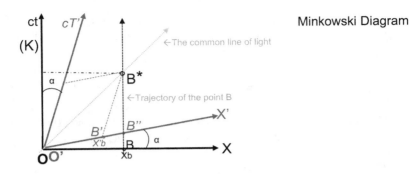

FIGURE 9.3
Minkowski diagram of simultaneous light pulses

Q2A: On the ground, the distance to the tunnel is given by the point B of OX having the abscissa Xb of the event B*.

Q2B: Seen from the train, the distance covered by the pulse is the segment OB' with B' being the point of O'X' having the same abscissa X'b as the event B*.

Q3A: OB is the proper distance from the point O to the tunnel because the tunnel is fixed relative to the ground.

Q3B: No, because seen from the train, the tunnel is moving closer and closer to the train, so that during the pulse trip to the tunnel, the tunnel will have moved toward the train. Hence, the distance train-tunnel at the time $t' = 0$ is longer than the distance covered by the pulse.

Q4: The distance train-tunnel seen from the train at the time $t' = 0$ is given by the point B'' which is on the intersection of the O'X' axis (since $t' = 0$) and the worldline of the point B of the tunnel. The tunnel being fixed on the ground, its worldline is a parallel line to OcT passing by B.

Q5: We notice that O'B'' is greater than O'B'. This is consistent with the fact that the light pulse emitted from the train covers a shorter distance than the distance train-tunnel at $t' = 0$, as expected from Q3B above.

9.3 Answers to Questions and Problems of Chapter 3

Problem 3.1: Relativity of present, past and future; impossibility of greater speeds than c

Q1A: Consider a frame K' which shares the common origin-event E* with K, and which is moving at the speed v relative

to K. The Lorentz transformation gives the time of F* in K':
$T'_f = Y(T_f - vX_f /c^2) = -Y \ vX_f/c^2$ (since $T_f = 0$, being equal in K to
$T_e = 0$). Hence, if v is positive, T'_f is negative, meaning that F* is
seen occurring before E* in K'. Conversely, if v is negative, the
event F* is seen occurring after the event E*.

Q1B: F* is seen prior to E* in any frame which move in the direction
EF. Conversely, F* occurs after E* in any frame moving in the
opposite direction.

Q2: We are in the case of negative v; hence, the previous question
showed that F* is seen after E* in K'. The distance covered by
the object M in K' is obtained by the Lorentz transformation:
$X'f = Y (X_f - vT_f) = Y X_f = Y.1$ m.

The time in K' taken by this object M from E* to F* is: T'_f, and we
saw that $T'_f = Y \ vX_f /c^2$. Consequently, $s = c^2. Y/Y v = c^2/v$. Then, as
$v < c$, the object speed s is greater than c.

Q3: In K" which is going at the speed v along the direction EF, the
object M is seen arriving in F* before leaving E* since the speed v
is in the direction of EF in K. This situation is absurd, which means
that there cannot be such an object M going at a faster speed than c.

Problem 3.2: The special inertial frame associated with a pair of events

Q1A: EF being a time-like interval and E the origin-event, the event
F is above the line of light. Hence, let's choose the frame K' such
that the point F is on its O'cT' axis. Indeed in K', the event F has
its abscissa null: it is co-located with E (Figure 9.4).

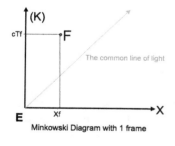
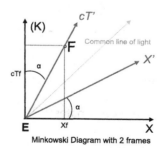

FIGURE 9.4
Existence of a frame where the 2 events are co-located

Q1B: K' is such that: $tg(\alpha) = \dfrac{cTf - cTe}{Xf - Xe}$.

Q2A: EF being a time-like interval and E the origin event, the event F
is below the line of light. Hence, let's choose the frame K" so that
the point F is on its O'X' axis. In K", the event F has its ordinate
null: it is simultaneous with E (Figure 9.5).

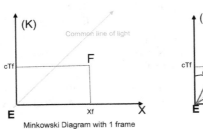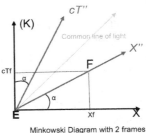

Minkowski Diagram with 1 frame   Minkowski Diagram with 2 frames

FIGURE 9.5
Existence of a frame where the 2 events are simultaneous.

Q2B: Again, $tg(\alpha) = \dfrac{cTf - cTe}{Xf - Xe}$.

Problem 3.3: Twin paradox with the simultaneity lines, Part 1

Q1: Alice's frame: K. Bob's frame: K'. The angle α $(OX, OX') =$ arctan0.866 = 40.9°; OF* = 10 years. The coordinates of F* in K' are: $(X' = 0; cT' = 10\,\text{years})$ (Figure 9.6).

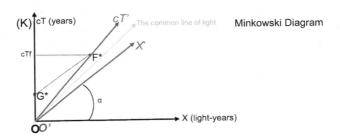

FIGURE 9.6
Minkowski diagram showing the line of simultaneity.

The coordinates of F*in K are: Xf = 10 × 0.866 = 8,66 light-years. Tf = Υ·10 = 20 years, since Alice sees the proper time duration of Bob's trip from O to F* dilated.

Q2A: The event G* being simultaneous with F* in Bob's frame, it is on the intersection between the axis O-cT (on which Alice is), and the parallel to O'X' passing by F*.

Q2B: Let Tg be the time in K of G*. The time Tg is Alice's proper time; hence, it is seen dilated by Bob, so:

Tg.Υ = Tf; then: Tg = 10/Υ = 5 years.

Q2C: We previously saw that Tf = 10.Υ and Tg = 10/Υ. Hence: Tf/Tg = γ^2.

Q3: The result of Q2B induces Bob to think that Alice will be younger than him at his return, but this is the opposite of the theoretical result. The reason for this discrepancy is that he will inevitably incur an acceleration phase before meeting Alice, during which he cannot apply the same reasoning as in Q2. Hence, he cannot extrapolate his reasoning to his entire trip.

Problem 3.4: Twin paradox with the simultaneity lines, Part 2

Q1A: Bob's new trajectory is a straight line passing by F* and reaching the axis OcT at the ordinate of M*, which is $2Tf = 40\,years$, because Bob goes at the same speed in his way back as in his way from O to F*.

Q1B: The new line of light starts at F* and goes perpendicularly to the first common line of light because its equation in K is: $(X - Xf) = -c\,(T - Tf)$ (Figure 9.7).

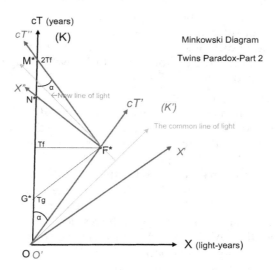

FIGURE 9.7
Minkowski diagram showing the lines of simultaneity at F* and just after F*.

Q1C: According to the Minkowski diagram principle, the new line of light is the bisector between the F*cT″ axis and the F*X″ axis.

Q2A: The event N* is at the intersection between F*X″ and OcT.

Q2B: Given the symmetry of the diagram, the classical length of the segment M*N* is equal to OG*. We saw that $Tg = 5\,years$ and $Tm = 40\,years$; hence, $Tn = 40-5 = 35\,years$.

We can thus see that Alice's age, seen by Bob, has dramatically changed after his U-turn, despite the fact that his U-turn was instantaneous. Before the U-turn, Bob considered Alice has spent 5 years, whereas after the U-turn, she has spent 35 years. The instantaneous U-turn thus caused 30 years change in Bob's perception of Alice's aging!

Q3A: The travel time seen by Bob for his return is the same as for his way from O to F*: 10 years. Hence, Bob calculates that his total trip will take 20 years. After the U-turn, Bob thinks that Alice will be older than him when he returns since he calculated that Alice's clock marked 35 years just after his U-turn.

Q3B: For Alice, Bob's proper time is seen dilated by the same gamma factor of 2; hence, she calculates that the return trip will take 20 years. Thus, for the whole trip, Bob will have spent 20 years, whereas Alice 40 years. She will thus be older than Bob at his return, in accordance with the theory.

Q3C: One cannot consider that Bob's proper time has flowed more slowly because he is younger at the arrival, because this reasoning would be false on the first part of the trip: Bob indeed considered that Alice was aging more slowly than him. Their aging difference is due to a continuous time de-synchronization between Alice's and Bob's clocks, with a huge de-synchronization gap during the U-turn.

Q4: The new diagram is the same as previously after swapping Bob's and Alice's names. Hence, the result is: Bob will have spent 40 years, whereas Alice 20 years. This result is in contradiction with the previous one, and this contradiction is a real one, since it concerns an intrinsic notion: the proper time (of Bob, and of Alice). It is not like the contradiction encountered with Einstein's train scenario, which concerned a non-intrinsic notion: the time of an inertial frame (cf. more in Volume II Section 1.3). Moreover, Bob and Alice's scenario can be made for real, with the result in both cases that Bob is younger than Alice when they meet again. The contradiction here actually does not exist because Bob used a Minkowski diagram although he was not justified since he is not in an inertial frame. (Laws of Special Relativity are imbedded in Minkowski diagram.)

Problem 3.5: Demonstration of the Lorentz transformation with natural units

Q1A: The coordinates of E* in K are in classical units: ($t_e = 1$ second; $x_e = 0$). In natural units, the coordinates of E* are: ($T_e = 1.c$ with $c \approx 3 \cdot 10^8$ m/s; and $X_e = 0$).

Q1B: The second flash takes place in O of K at the time 1 second; hence, it is seen dilated in the frame K'. We thus have with classical units: $t'_e = \gamma_v t_e$.

This expression becomes in natural units: $ct'_e = \gamma_\beta.c\, t_e$ with $\beta = v/c$. Then: $T'_e = \gamma_\beta.T_e$, and so: $T'_e = \gamma_\beta.c$.

Regarding X'_e: seen from K', the point O has moved in the opposite direction to $\vec{v}$, and for a time t'_e. Hence: $x'_e = -v.t'_e$. Then in
natural units: $X'_e = x'_e = -\beta.T'_e = -\beta.\gamma_\beta.c$.

Q1C: The equation $X'_e = aX_e + bT_e$ becomes: $-\beta.\gamma_\beta.c = b.c$, so: $b = -\beta.\gamma_\beta$.
The equation $T'_e = mX_e + nT_e$ becomes: $\gamma_\beta.c = n.c$, so: $n = \gamma_\beta$.

Q2A: In classical units the coordinates of F* in K' are ($t'_f = 1$ second;
$x'_f = 0$). In natural units, we have: $T'_f = 1.c$ (with $c \approx 3 \cdot 10^8$ m/s) and
$X'_f = 0$.

Q2B: In K, in classical units, the time $t'_f = 1$ second is seen dilated
by γ_v. So: $t_f = \gamma_v . t'_f$. In natural units, the same reasoning leads to:
$T_f = \gamma_\beta.T'_f = \gamma_\beta.c$.

Regarding X_f, let's reason directly in natural units: seen from
K, the point O' has moved in the direction of $\vec{\beta}$ during T_f. Hence:
$X_f = \beta.\gamma_\beta.c$.

Q2C: $X'_f = aX_f + bT_f$ becomes: $0 = a.\beta.\gamma_\beta.c + b.\gamma_\beta.c$, then: $a.\beta = -b = +\beta.\gamma_\beta$,
so: $a = \gamma_\beta$.

$T'_f = mX_f + nT_f$ becomes: $c = m.\beta.\gamma_\beta.c + n.\gamma_\beta.c$, then: $1 = \gamma_\beta.(m\,\beta + \gamma_\beta)$.

Then: $m\beta = \dfrac{1}{\gamma_\beta} - \gamma_\beta = \dfrac{1 - \gamma_\beta^2}{\gamma_\beta} = \dfrac{-\beta^2}{(1-\beta^2)\gamma_\beta} = \dfrac{-\beta^2.\gamma_\beta^2}{\gamma_\beta}$, so finally:

$m = -\beta.\,\gamma_\beta$.

The   Lorentz   transformation   thus   reads:

$$\begin{pmatrix} X' \\ T' \end{pmatrix} = \gamma_\beta \begin{pmatrix} 1 & -\beta \\ -\beta & 1 \end{pmatrix}\begin{pmatrix} X \\ T \end{pmatrix}.$$

Problem 3.6: Synchronization impossibility by moving a clock

Q1A: Let's call F* the event: the traveling clock C is in front of B. In K,
F* has for coordinates: (AB, AB/v). The time coordinate of F* in
K', denoted by T'_f, can be obtained by the Lorentz transformation:
$T'_f = \gamma.(AB/v - v/c^2AB) = \gamma.AB.(1 - \beta^2)/v = AB/v.\gamma = T_f/\gamma$. This
matches with the time dilatation law: the time T_f taken by the
moving clock is seen dilated in K: $T_f = \gamma. T'_f$.

Q1B: T'_f being less than T_f, the clock C displays at the point B a time
which is behind the clock B, which displays T_f since this latter
clock is synchronized with the time of K.

Q2A: Langevin's twin Travelers paradox shows that the accelerated
clock D is behind the inertial clock C.

Q2B: The first answer showed that the inertial clock C, when arriving at B, displays a time which is behind that of the frame K. The
second answer then implies that the accelerated clock D displays
a time which is behind the time of the frame K.

Problem 3.7: The clock in a satellite

The satellite having a circular trajectory, it has an acceleration (centripetal). Nevertheless, the Earth frame being inertial, observers on Earth can apply the laws of Special Relativity on small segments of the satellite trajectory. Consequently, the satellite speed relative to the Earth generates a time dilatation effect: the satellite clock will be seen on Earth beating more slowly.

The satellite travels $(106{,}000 \times 2 \times 3.14)$ km in $1\ h = 665{,}680$ km/h; hence, $\Upsilon = 1.00000019$. The satellite clock is seen running more slowly by: $(1.00000019 - 1) \times 100\% = 0.000019\%$.

Problem 3.8: Two opposite light beams

The segment AB is increasing at the speed 2 c. This is not in contradiction with c being the maximum possible speed, because the rate of increase in the segment AB does not correspond to the speed of any single object relative to an inertial frame. In particular, this double beam mechanism cannot carry information from A to B.

Problem 3.9: Reflection of a beam on a mirror

Q1: In K′, the component along O′X′ of the velocity of the incident ray (A′B′) is: $w'_x = w_x \oplus (-v)$. Still in K′, the reflected ray (B′C′) has the same angle with the mirror as the incident ray, denoted by θ. The horizontal component of the speed of the reflected ray B′C′ is

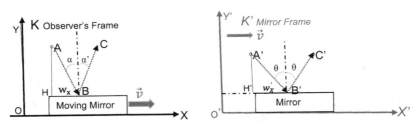

FIGURE 9.8
Light beam hits a moving mirror.

then equal to w'_x (Figure 9.8).

Subsequently, in K, let's call w the horizontal component of the speed of the reflected beam (BC); we have: $w = w'_x \oplus v = w_x \oplus (-v) \oplus v = w_x$. Thus, the horizontal component of the speed of the reflected beam (BC) is the same as that of the incident beam. Since both beams move at the speed c, we have: $w_x = c\sin\alpha = w = c\sin\alpha'$. Hence, $\alpha' = \alpha$.

Q2: In K′, we have: $\sin\theta = \dfrac{w'_x}{c}$ and $w'_x = w_x \oplus (-v) = \dfrac{w_x - v}{1 - vw_{x/c^2}}$.

Hence: $\sin\theta = \dfrac{\sin\alpha - \beta}{1 - \beta\sin\alpha}$, with $\beta = v/c$.

Problem 3.10: Velocity composition and the triangle inequality

Q1: The velocity $\overrightarrow{W_1}$ being parallel to the motion of K' relative to K, we have:

$$w_1 = \frac{s + v}{1 + sv/c^2} = \frac{\dfrac{c}{2} + \dfrac{3c}{4}}{1 + 3c^2/8c^2} = \frac{5c/4}{1 + 3/8} = \frac{1.25\,c}{1.375} \approx 0.91\ c.$$

Comment: It is smaller than $s + v = 1.25\ c$, and thus prevents the resulting speed from being greater than c.

Q2: The velocity of $\overrightarrow{W_2}$ being perpendicular to the motion of K' relative to K, we have:

$$w_{2y} = \frac{s}{\gamma} = \frac{c/2}{1.51} = 0.33\ c \quad \text{and} \quad w_{2x} = \frac{0 + v}{1 + 0.v/c^2} = v = 0.75\ c.$$

Q3: In the frame K', the mobile object covers a distance which is less than the first object by the factor: $\dfrac{\sqrt[2]{2}}{2}$. Its speed is then:

$$W_3' = s.\ \frac{\sqrt[2]{2}}{2} = c.\ \frac{\sqrt[2]{2}}{4} \approx 0.35c.$$

The component of W_3' which is parallel to OX is equal to:

$$W_{3x}' = s.\ \frac{\sqrt[2]{2}}{2} = c\frac{\sqrt[2]{2}}{4}.\frac{\sqrt[2]{2}}{2} = c/4.$$

The component of W_3' which is parallel to OY is also equal to:

$$W_{3y}' = s.\ \frac{\sqrt[2]{2}}{2} = c/4.$$

Q4: The speed W_{3x} being parallel to the motion of K' relative to K, we have:

$$W_{3x} = \frac{W_{3x}' + v}{1 + W_{3x}'v/c^2} = \frac{c.\dfrac{1}{8} + c.\dfrac{3}{4}}{1 + 3c^2/16c^2} = \frac{c.\dfrac{7}{8}}{19/16}\quad 0.84\ c.$$

Q5: The speed w_{3y} being perpendicular to the motion of K' relative to K, we have:

$$W_{3x} = \frac{W_{3y}'}{\gamma\left(1 + W_{3x}'v/c^2\right)} = \frac{c/8}{1.51.\ \times\ 19/16}\quad 0.139\ c.$$

Q6A: In K', both objects take the same time according to the problem statement, and it is: $2\times AB/s = 4\,\text{years}$.

Q6B: In K, both objects also take the same time because of event coincidence invariance: if both objects are seen arriving simultaneously at C in one frame, they are also seen arriving simultaneously at in all frames (cf. Volume II, section 1.3.1). The event coincidence invariance implies that there is a one-to-one relationship between an event and its set of four coordinates in any inertial frame.

Q6C: In K, the distance AB is seen contracted by $\gamma = 1.51$; hence, it is: $1/1.51 = 0.66$ light-years. The distance BC is invariant, being transverse; but both B and C are moving, so we must use the transverse component of w_2. Hence, the time taken in K by both objects is:

$$0.66 / w_1 + 1 / w_{2y} = 0.66 / 0.91 + 1 / 0.33 = 0.72 + 3.02 = 3.74 \text{ years}$$

Q7A: The proper time of the second object is seen dilated in K' by the factor $\gamma_{w_3'} = \gamma_{0.35c} \approx 1.069$. We then have: $\tau_2 = 4/1.069 = 3.75 \text{ years}$.

Q7B: The proper time of the first object is less because the instantaneous turn in B causes its trajectory to be non-inertial from A to C; and the ds^2 triangle inequality shows that:

$$\sqrt{ds^2(AC)} > \sqrt{ds^2(AB)} + \sqrt{ds^2(BC)}.$$

Besides, we have: $c\tau_1 = c.\sqrt{ds^2(AB)} + c.\sqrt{ds^2(BC)}$.

Q7C: The proper time duration of the first object between A and B. This object sees the segment AB contracted; hence, it is 1 light-year/ $\gamma_{C/2} = 1/1.155 = 0.87 \text{ years}$. The same applies to the segment BC, so the proper time duration of the first object going from A to C is $\tau_1 = 1.94 \text{ years}$, which indeed is less than τ_2.

9.4 Answers to Questions and Problems of Chapter 4

Question 4.1: No, it is not necessary. It is done with the purpose of having the same unit for the relativistic Momentum as for the classical one: mass × speed.

Question 4.2: With fully natural units, the time and the distance are always expressed with a single unit. It is not the case with the homogeneous (natural) units since the numerator is expressed with a unit of distance (meter), whereas the denominator with a unit of time (second). Conversely, with fully natural unit the denominator is a time also expressed with the distance unit $t \rightarrow ct$. (For example, the speed β is a number without dimension.)

Question 4.3: Heavier in summer because the higher the temperature, the greater the speeds of the water molecules; hence, the greater is the system energy in the CoM frame, $\Sigma mc\gamma_i$, which is equal to the system mass/c^2.

Question 4.4: The object velocity is always null in its own frame; hence, $P = (mc, 0)$.

Problems regarding Classical Physics and Relativity

Problem 4.1: Combination of inertial and gravitational forces

Q1: My weight having decreased, I can deduce that I am subject to an inertial force which is upward. The inertial force is always opposite to the frame acceleration relative to an inertial frame; hence, the lift acceleration must be downward.

Q2: The force F that I exert on the scale for weighing inside the lift is: $F = W + F_i$, with W being my weight in an inertial frame, and F_i the inertial force: $F_i = -m.a$, with a being the lift acceleration. We then have: $90\%.mg = mg - ma$. So: $a = 0.1g = 0.980\,\text{m/s}^2$.

Problem 4.2: Altitude of a geostationary satellite

The satellite incurs a centrifugal force of inertia, F_i, since its trajectory it is not inertial, being circular. This force offsets the gravitational force. We denote by R the Earth radius, and by H the satellite altitude above the Earth surface. We have: $F_i = m.\omega^2.(R + H)$, with ω being the satellite angular speed:
$\omega = 2\pi/(24 \times 3600) = 7.27.\,10^{-5}$ rad/s. The gravitational force is: $mMG/(R+H)^2$. Consequently: $mMG/(R + H)^2 = m.\omega^2.(R+H)$;
then: $(R+H)^3 = [(5.972 \times 10^{24}\,\text{kg}).(6.674 \times 10^{-11}\,\text{m}^3/\text{kg.s}^2)]/(7.27 \cdot 10^{-5})^2$.
With $R = 6{,}400\,\text{km}$, we find: **H = 35,784 km**. Remark: The satellite mass m is irrelevant: the object trajectory is independent of its mass, which is consistent with the equality between the inertial mass (involved in the inertial force), and the gravitational mass.

Problem 4.3: Keplerian orbital law

Q1A: The planet incurs a centrifugal force that offsets the force of gravity: $mV^2/R = \dfrac{mMG}{R^2}$, so: $V = \sqrt{\dfrac{MG}{R}}$. Thus, the further a planet is from its star, the slower its speed.

Q1B: Having $V = \omega R$, the previous reasoning yields: $m\omega^2 R = \dfrac{mMG}{R^2}$;

hence: $\omega = \sqrt{\dfrac{MG}{R^3}}$.

Q2A: From Q1A, : $\dfrac{\omega_{Mercury}}{\omega_{Earth}} = \left(\dfrac{R_{Earth}}{R_{mercury}}\right)^{3/2} = \left(\dfrac{1}{0.47}\right)^{3/2} = 2.127 \times 1.458 = 3.195.$

Hence, Mercury takes $365/3.195 = 114$ days to make 1 revolution around the Sun. (This is an approximation since the actual trajectory is elliptic.)

Q2B: From Q1A, we have: $\dfrac{V_{Mercury}}{V_{Earth}} = \left(\dfrac{R_{Earth}}{R_{mercury}}\right)^{1/2} = \left(\dfrac{1}{0.47}\right)^{1/2} = 1.46;$ hence,

$V_M \approx 156\ 200$ km/h.

Problem 4.4: Weight is not only gravity

Q1: Greater, since one incurs at the equator a centrifugal force, F_i, due to the rotation of the Earth which is in the opposite direction from gravitation. This centrifugal force is null at the Pole.

Q2: The gravitational force is: $m.g$ and $F_i = m\ \omega^2 R$. So: $F_i = \omega^2 R = \left(\dfrac{2\pi}{24 \times 60 \times 60}\right)^2 \times 6{,}371 = 0.034\,\text{ms}^{-2}.$ Having $g = 9.81\,\text{ms}^{-2}$, the weight at the Pole is greater by 0.35%, meaning 243 g.

Q3: At the Pole, the radius reduces by $dR/R = 0.35\%$. Hence, gravitation increases by $\approx 2dR/R = 0.7\%$, meaning 490 g. In all, the person weighs 733 g more at the Pole than at the equator, meaning 1.05%.

Problems regarding Relativity

Problem 4.5: An inelastic collision

Q1: Before the collision, the classical momentum of A is: $2 \times 1V = 2V$; that of B is: $1 \times (-2V) = -2V$. The system momentum is then null, and its conservation implies that C will be fixed in K.

Q2: In Relativity, the spatial part of the Momentum of A is $P_S^A = 2V\,\gamma_V$; that of B is: $P_S^B = -2V\,\gamma_{2V}$. We then have: $P_S^B > P_S^A$, which implies that the object C will move in the same direction as B initially.

Q3: We have no guarantee that the sum of the masses is conserved, and we will see with Problems 5–6 that it is not conserved.

Problem 4.6: The contravariance property 1

Q1: The Momentum in K is: $(2c, 0)$. The Momentum in K′ is obtained from that of K by applying the Lorentz transformation: $\vec{P'} = (\Lambda)\ \vec{P}.$

With the matrix format: $\begin{pmatrix} P'_t \\ P'_x \end{pmatrix} = 1.40 \begin{pmatrix} 1 & -0.7 \\ -0.7 & 1 \end{pmatrix} \begin{pmatrix} 2c \\ 0 \end{pmatrix}.$

Hence:

$P'_t = 1.40 \times (2 \times c - 0.7 \times 0) = 2.8$ c; $P'_x = 1.40 \times (-2 \times c \times 0.7 + 1 \times 0) = -1.96$ c.

Q2: In K', the object velocity is: $-0.7c$. Then $P'_t = mc \, \gamma = 2.c.1.40 = 2.80$, and $P_s = -2 \times 0.7c \times 1.40 = -1.96$.

Q3: In K is $P^2 = (2c)^2 - (0)^2 = 4 \, c^2$. In K': $P^2 = (2.8^2 - 1.96^2)c^2 = 3.9984c^2$.

Problem 4.7: The contravariance property 2

Q1: The Momentum in K is:

$(2c \times 1.048, 2 \times 1.048 \times 0.3 \, c) = (2.096 \, c, 0.6288 \, c)$.

Q2: The Momentum in K' is obtained from that of K by applying the Lorentz transformation: $\vec{P'} = (\Lambda) \, \vec{P}$.

With the matrix format: $\begin{pmatrix} P'_t \\ P'_x \end{pmatrix} = \gamma \begin{pmatrix} 1 & -\beta \\ -\beta & 1 \end{pmatrix} \begin{pmatrix} P_t \\ P_x \end{pmatrix}$; then

$\begin{pmatrix} P'_t \\ P'_x \end{pmatrix} = 1.40 \begin{pmatrix} 1 & -0.7 \\ -0.7 & 1 \end{pmatrix} \begin{pmatrix} 2.096c \\ 0.6288c \end{pmatrix}$.

Hence: $P'_t = 1.40 \times (2.096 \times c - 0.7 \times 0.6288 \, c) = 2.318$ c.

$P'_x = 1.40 \times (-2.096 \times c \times 0.7 + 1 \times 0.6288 \, c) = -1.174$ c.

Then in K: $P^2 = (2.096c)^2 - (0.6288c)^2 = 3.99 \, c^2$.

In K', $P^2 = (2.318 \, c)^2 - (-1.174 \, c)^2 = 3.99$.

Problem 4.8: The Momentum density defined with the mass density

Q1: The 4D Vector $\vec{D}$ is contravariant because it is proportional to the proper speed $\vec{S}$, which is contravariant, and the proportionally coefficient is invariant.

Q2: In K, with homogeneous units, we have:

$\vec{D} = \rho_0 \vec{S} = (\rho_0.c.\gamma_v, \, \rho_0.\gamma_v.\vec{v})$.

Q3: In K, the length of the cube in the direction of the motion is seen contracted by γ_v. The dimensions of the other sides are unchanged, being along transverse directions. Hence, the cube volume is seen reduced in K by γ_v, so its density is increased by γ_v. Thus: $\rho = \gamma_v \rho_0$.

Q4: From 2 and 3, we have in K: $\vec{D} = \rho_0 \, \vec{S} = (\rho.c \, , \, \rho.\vec{v} \,)$.

Q5: No, because in $K°$, $\overline{E^0} = \rho_0 \, \overline{S^0} = \overline{D^0} = \rho_0(c, 0)$. Then if we perform the Lorentz transformation of E^0 from $K°$ to K, we obtain: $(\Lambda)\rho_0 S^0 = \rho_0 \vec{S}$, whereas $\vec{E} = \rho \vec{S}$ in K.

9.5 Answers to Questions and Problems of Chapter 5

Question 5.1: $m < m_A + m_B$ (cf. §5.2.3).

Question 5.2: With fully natural units, $E = M$; hence, the MeV (which is a unit of energy) can also be a unit of mass. However, in the international system of unit, their common unit it the kg.

Question 5.3: The two Forces are identical in the object's Inertial Tangent Frame by definition.

*

Problem 5.1: Natural units
$105 \, \text{GeV} \rightarrow 105 \times 10^9 \times 1.602 \times 10^{-19} \text{joule/eV}/9 \times \ 10^{16} = 1.87 \times 10^{-26} \text{kg}$.

Problem 5.2: The deuteron
Let's compare its mass with the sum of its nucleons: proton: 1.00727 u + neutron: 1.00867 u = 2.01594 u.
We have: 2.01355 u − 2.01594 u = −0.00239 u, which represents 0.12% of the deuteron mass. Hence, the deuteron is stable as its disintegration requires an energy corresponding to the mass of 0.00239 u, which is: 0.00239 u × 0.9315 (GeV/c² per u) = 2.23 MeV.

Problem 5.3: Hydrogen ionization
The Hydrogen atom mass before ionization has a mass defect corresponding to the binding energy of the electron, which is 13.58 eV. Hence after ionization, the atom mass will increase by 13.58 eV/c², which is in atomic unit mass; 13.58 eV /0.9315 GeV = 14.9 × 10⁻³ u. Then, in kg: $14.9 \times 10^{-3} \text{u} \times 1.66 \times 10^{-27}$ u/kg = 24,7 × 10⁻³⁰/ kg.

Problem 5.4: The mass of two photons going in opposite directions is NOT null
The system Momentum is constant:
$$\vec{P}_{Tot} = (E/c,\vec{p}) + (E/c, -\vec{p}) = (2E/c, \vec{0}) = (2h\nu/c, \vec{0}).$$

The system mass is: $M_{tot}^2 c^2 = P_{tot}^2 = 4h^2\nu^2 - 0$. Consequently, the system mass of these two photons is: $M_{tot} = 2h\nu/c$, which may seem surprising because each photon' mass is null.

Problem 5.5: The relativistic Momentum solves the inelastic collision Problem Section 4.1.2

Q1: The space part of the system Momentum before collision is null for symmetry reasons. After collision, it is also null because the motion is null. The system Momentum conservation is verified.

Q2A: In K', the spatial part of the system Momentum in K' after collision is: $SM'_2 = -2m'V.\gamma_v$.

Q2B: In K', before the collision, the spatial part of the system Momentum, denoted by $SM1'$, is:

$$SM'_1 = m.V'_{A1}\gamma_{V'_{A1}} + m.V'_{B1}\gamma_{V'_{B1}} = 0 - m.(V \oplus V)\gamma_{v\oplus v}, \text{ with VA1' and}$$

VB1' being the speeds of A and B seen in K' before the collision. The square of the right part is:

$$m^2 \frac{4V^2}{(1+\beta^2)^2} \frac{1}{(1-\dfrac{4\beta^2}{(1+\beta^2)^2})} = m^2 \frac{4V^2}{(1+\beta^2)^2} \frac{(1+\beta^2)^2}{((1+\beta^2)^2 - 4\beta^2)}$$

$$= m^2 \frac{4V^2}{(1-\beta^2)^2} = m^2 4V^2.\gamma_v^4.$$

Thus, the spatial part of the system Momentum in K' before collision is: $SM'_1 = -2mV\gamma_v^2$.

Q2C: If $m' = m$, then the system Momentum conservation law is not respected in K' since the previous results give: $S M'_1 = S M'_2/\gamma_v$. The condition for this law to be respected is $m' = m. \gamma_v$.

Q3A: In K, the total energy in natural units is the sum of the inertia.
  - Before the collision, it is $E_1 = 2m \gamma_v$. After the collision, it is $E_2 = 2m'$ (since $\gamma_\circ = 1$).
  - The system energy conservation law gives: $2m \gamma_v = 2m'$. Hence, we again obtain: $m' = m \gamma_v$.

Q3B: In K', after the collision, the system energy is: $E'_2 = 2m' \gamma_v$.

  - Before the collision, the system energy is:

$$E'_1 = m + m.\gamma_{v\oplus v} = m + m \gamma^2_v (1 + \beta^2) = m(\frac{1-\beta^2 + 1 + \beta^2}{1-\beta^2}) = 2m. \gamma^2_v.$$

  - The system energy conservation law then gives:

$2m' \gamma_v = 2m. \gamma^2_v$, so again: $m' = m\gamma_v$.

Q4A: In K, the kinetic energy before collision in natural units is:
$E_1 - 2m = 2m . (\gamma_v - 1)$. After collision, it is: 0.

Q4B: In K', after collision, it is: $E'_2 - 2m' = 2m'(\gamma_v - 1) = 2m \gamma_v (\gamma_v - 1)$.
Before collision, it is: $E'_1 - 2m = 2m. (\gamma^2_v - 1)$. The kinetic energy variation is: $-2m(\gamma_v - 1)$, which is the same as in the frame K. It is

normal because it corresponds to the mass variation (in natural units), and the mass is an intrinsic notion. In other words, the kinetic energy loss due to the collision has been transferred to the objects' masses.

Q5A: The system mass², denoted by M^2, is equal to $P^2 = ds^2(P)$ with natural units. We notice that K is the Center of Momentum frame; hence, the system mass is the sum of the inertias in K: $M = 2m\gamma_v$ before the collision; after the collision, it is $2m'$. The constancy of the system mass of an isolated system requires that $m' = m\gamma_v$, and we can see that this condition is satisfied.

Q5B: In K', the system mass before the collision is with natural units:
$$P^2 = E_1'^2 - SM_1'^2 = 4m^2\gamma_v^4 - 4m^2\beta^2\gamma_v^4 = 4m^2\gamma_v^4(1-\beta^2) = 4m^2\gamma_v^2,$$
hence: $M = 2m\gamma_v$.
The system mass after the collision is with natural units:
$$P^2 = E_2'^2 - SM_2'^2 = 4m'^2\gamma_v^2 - 4m'^2\beta^2\gamma_v^4 = 4m'^2\gamma_v^4(1-\beta^2) = 4m'^2\gamma_v^2;$$
hence, $M = 2m'\gamma_v$. We can thus see that with $m' = m\gamma_{v.}$, **the system mass is constant and invariant** (meaning that it doesn't vary with time, and is identical in all inertial frames).

Problem 5.6: An inelastic collision

The system Momentum conservation yields:
$$P_{tot} = (2c1.067 + 1c1.4, 2 \times 0.35\ c \times 1.067 - 1 \times 0.7\ c \times 1.4)$$
$$= (3.534\ c, -0.2331\ c).$$

Also, $P_{tot} = (Mc\gamma, Mv\ \gamma)$. The mass is given by:
$$P_{Tot}^2 = M^2c^2 = (3.534c)^2 - (0.2331c)^2 = 12.43c^2.$$

Thus, $M = 3.53\,kg$. We can see that it is greater than the sum of the initial masses.

The speed of C is given by: $\beta = \dfrac{P}{E} = \dfrac{Mv\gamma}{Mc\gamma} = \dfrac{-0.2331c}{3.534c} = -0.66.$

The minus sign means that C goes in the same direction as B.

Problem 5.7: Example of mass changes after an inelastic collision

The system energy conservation yields: $2m\gamma = 2m'$. So $m' = 1{,}000 \times (1 + 5.66 \times 10^{-10}) = 1{,}000.000000566\,kg$. Such a weight increase can be detected, but the speed assumption of 36,000 km/h is extremely high.

Problem 5.8: Atom emitting a photon

Q1: The reaction being spontaneous, the system Momentum is conserved: $(m, 0) = (m'.\beta.\gamma) + (\varphi, \varphi)$. Then: $m = m'\ \gamma + \varphi$ (1)
and $m'\beta\ \gamma = \varphi$ (2). Hence, from (1), we deduce: $m' < m$, and from (2), the atom acquires the speed β which is $\neq 0$.

Q2: From (2), we have:

$m'\beta\,(1+\beta^2/2) \approx -\varphi$ (3). From (1), we have: $m'(1+\beta^2/2) \approx m - \varphi$. (4).

Let's divide (3) by (4): $\beta = \dfrac{-\varphi}{m-\varphi} \approx -\dfrac{\varphi}{m}\,(1+\varphi/m)$ (5). Then,

with usual units: $v/c = -\dfrac{\varphi}{mc^2}\,(1+\varphi/mc^2)$ (6).

Then from (4): $m' = \dfrac{m-\varphi}{\left(1+\dfrac{\varphi^2\left(1+\dfrac{\varphi}{m}\right)\varphi}{2m^2}\right)} \approx (m-\varphi)\left(1-\dfrac{\varphi^2}{2m^2}\right) \approx$

$m-\varphi-\dfrac{\varphi^2}{2m}$. Then: $\mathbf{m-m'} \approx \boldsymbol{\varphi} + \dfrac{\varphi^2}{2m}$.

Converting into usual units: $-dm.c^2 = \varphi + \dfrac{\varphi^2}{2mc^2}$ (7).

The atom energy variation is: $m-m'\,\gamma \approx m - m\,\gamma - \varphi\,\gamma - \dfrac{\varphi^2\gamma}{2m}$

$\approx -(\varphi+\dfrac{\varphi^2}{2m})$. With usual units: $\Delta E = -(\varphi+\dfrac{\varphi^2}{2c^2m})$.

Q3: We have: $\dfrac{\varphi}{mc^2} \approx \dfrac{3.90\ \text{MeV}}{12.u.0.9315\dfrac{\text{GeV}}{c^2 \times c^2}} \approx 3.5\times10^{-4}$.

Then, from (6): $\beta = 3.5\times10^{-4}\,(1+3.5\times10^{-4}) \approx 3.5\times10^{-4}$, which is low enough to justify the approximation with ($\gamma \approx 1+\beta^2/2$). So: $v \approx 105$ km/s.

We have: $m = m'\,\gamma + \varphi$; hence:

$m = 12\times0.9315$ GeV $\times 1.00000006125 + 3.90$ MeV $= 11.184$ GeV.

The final mass is: 12×0.9315 GeV $= 11.178$ GeV; hence, the mass loss is: 3.9000684 MeV, or 67.4 eV more than the photon energy.

We can see that the atom mass loss has been converted almost entirely into the photon energy, while a tiny fraction, 17.5 parts per million, goes into the kinetic energy acquired by the atom (recoil).

Problem 5.9: Ice melting and impact on the water mass

The system energy conservation yields:

10^6 kg.c$^2 + 10^6 \times 3.35\times10^5 = m'\,c^2$.

Then: $m' = 10^6 + \dfrac{3.35\times10^6\times10^5}{9\times10^{16}}$, and so $m' = 10^6$ kg $+ 3.7\times10^{-6}$ kg.

Problem 5.10: Objects' speeds after an elastic collision

In the CoM frame, the speeds of A and B are, respectively, Va and Vb before the collision, and V2a and V2b after the collision.

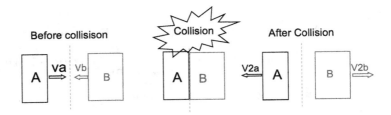

FIGURE 9.9
Elastic Collision.

The system Momentum conservation law yields in the CoM frame:

- for the spatial part: $M_a \vec{V_a}\gamma_{va} + M_b \vec{V_b}\gamma_{vb} = 0 = M_a \vec{V_{2a}}\gamma_{v2a} + M_b \vec{V_{2b}}\gamma_{v2b}$ (1)

- for the time part: $M_a c\gamma_{va} + M_b c\gamma_{vb} = M_a c\gamma_{v2a} + M_b c\gamma_{v2b}$. (2)

We took into account the constancy of the objects masses since the collision is elastic. These equations (1) and (2) enable us to calculate V2a and V2b knowing Va and Vb: In equation (1), one couple of solutions is obvious: $(\vec{V_{2a}} = -\vec{V_a})$ and $(\vec{V_{2b}} = -\vec{V_b})$ (3).Moreover, we can see that these solutions also satisfy (2). Consequently, these are the couple of solutions.

It still remains to determine the speed of B relative to A after the collision, and denoted by S2b:

Sb is the composition of the speed Vb with the speed

Va: Sb = Vb ⊕ Va.

Similarly, S2b is the composition of the speed V2b with V2a, but in the opposite direction as Sb; thus: S2b = −(V2b ⊕ V2a). Then with (3): S2b = −(Vb ⊕ Va) = −Sb.

This result shows that seen from the object A, the energy of the object B has not changed. However, we couldn't have deduced this result from the Momentum conservation applied from the frame of A because it is not inertial (due to the collision).

Problem 5.11: The speed such that the kinetic energy equals the energy due to the mass

Q1: The kinetic energy is $K = E - mc^2 = mc^2 (\gamma - 1)$. We must have $K = mc^2$; hence: $mc^2 = mc^2 (\gamma - 1)$, so: $\gamma = 2$.

Then: $4 = \dfrac{1}{1-\beta^2}$, and then $4\beta^2 = 3$, so: $\beta = \sqrt{0.75} = 0.866$ or v = 0.866 c.

Q2: $10 \times mc^2 = mc^2 (\gamma - 1)$, so $\gamma = 11$; hence: $121 = \dfrac{1}{1-\beta^2}$, and then $121\beta^2 = 120$, so: $\beta = \sqrt{\dfrac{120}{121}} = 0.9959$, or v = 0.99599 c. These particles are ultra-relativistic, and Problem 5.15 below addresses these cases.

Problem 5.12: Sun lifetime estimation

The yield of a thermonuclear reaction is approximately 0.7% (cf. Section 5.2.4). Hence, the 10% of the "burnable" Sun mass will create a quantity of radiated energy of:

$0.007 \times 0.1 \times 2 \times 10^{30} \times c^2 = 14 \times 9 \times 10^{30+16-4} = 1.3 \times 10^{44}$ joules.

Since $3.85.10^{26}$ joules is radiated every second, the duration for burning the total energy is:

$1.3 \times 10^{44} / 3.85 \times 10^{26} \approx 3.4 \times 10^{17}$ seconds. There are $\approx 3.15 \times 10^7$ seconds in 1 year, hence the number of years for the Sun to burn is ≈ 10.8 billion years.

Problem 5.13: The Carbon-12 binding energy

We have: proton mass = 1.00278 u, and neutron mass = 1.00867 u. Hence, we must provide mass/energy to break the carbon-12 into its nucleons, meaning that carbon-12 is stable.

The energy required to completely break up the nucleus is the binding energy, and its value for carbon-12 is:

$B = c^2 \times [6 \times (1.00278 + 1.00867) - 12] \times 0.9315 \text{ GeV/c}^2 = 89 \text{ MeV}$.

Note that it is important to distinguish nuclear masses from atomic weights, because 6×511 MeV is 3.07 MeV. The amu unit u (dalton) does include electron rest masses.

Problem 5.14: The decay of Beryllium

The binding energy is the difference between the sum of the nucleons and the Beryllium nucleus mass:

$6536 - 4 \times 938.272 - 3 \times 939.565 = 35.783$ MeV.

The system energy conservation gives: $6536 + 939.565 = 2 E_\alpha$. Then, $E_\alpha = 3737.7825$.

The kinetic energy is: $E_\alpha - 3728 = 9.78$ MeV.

Problem 5.15: Energy of ultra-relativistic particles

Q1: $\gamma = \sqrt{\dfrac{1}{1-\beta^2}} = \sqrt{\dfrac{1}{1-(1-\varepsilon)^2}} = \sqrt{\dfrac{1}{2\varepsilon - \varepsilon^2}} = \sqrt{\dfrac{1}{2\varepsilon(1-\varepsilon/2)}} \approx \sqrt{\dfrac{1}{2\varepsilon}}(1+\varepsilon/4) \approx \sqrt{\dfrac{1}{2\varepsilon}}$.

Then: $E = mc^2\gamma \approx \dfrac{mc^2}{\sqrt{2\varepsilon}}$.

Q2: We have $\varepsilon = 17.10^{-12}$, so

$E = \dfrac{mc^2}{\sqrt{2 \times 17.10^2}} = 1.71\text{x}10^5 mc^2 = 1.71\text{x}0.51\text{x}10^5 \text{MeV}$

$= 8.7.10^4 \text{MeV} = 87 \text{GeV}$.

The mass contribution to the total energy is then:

$r = \dfrac{mc^2}{E} = \dfrac{0.51\text{Mev}}{87\text{GeV}} = \dfrac{510\text{Kev}}{87\text{GeV}} = 5.9.10^{-6}$.

Problem 5.16: The acceleration due to the Newtonian-like Force

When the object speed is null, the spatial part of the Newtonian-like Force is identical to the classical force, since: $\vec{F}_s = m\gamma[\vec{a} + \vec{v}\gamma^2(\vec{\beta}\cdot\vec{\beta}')]$ with $\beta = 0$; hence, $a = F/m$.

When the speed is 0.9 c: First, the Force having been constant, the object is moving along a straight line parallel to this Force. Hence, we have:

$F = m\gamma a + m\gamma^3 0.9.(a/c).0.9c = ma(2.3 + 12.2 \times 0.81)$. So: $a = F/12.2\,m$.

We can thus see that for the same Force, the acceleration has been divided by 12.2.

Problem 5.17: The Fission Condition

The Binding energy of A is: $B(A) = AR$. The one of $(A_1 + A_2)$ is: $B(A_1 + A_2) = (A_1R_1 + A_2R_2) > (A_1 + A_2)R$. Knowing that: $A = A_1 + A_2$, we conclude: $B(A_1 + A_2) > B(A)$. This shows that the reaction leads to a greater binding energy, meaning a greater overall mass defect: it is then possible.

9.6 Answers to Questions and Problems of Chapter 6

Problem 6.1: Tidal effect from the Sun and from the Moon

Q1: The attraction from the Sun is:

$6.67 \times 10^{-11} \times 2 \times 10^{30} / 1.5^2 \times 10^{-22} = 5.93 \times 10^{-3}$ m/s².

The attraction from the Moon is:

$6.67 \times 10^{-11} \times 7.43 \times 10^{22} / 3.8^2 \times 10^{16} = 3.39 \times 10^{-5}$ m/s².

The attraction from the Sun is thus 175 times greater than the one by the Moon.

Q2: The tidal effect from the Moon is more important than the one from the Sun because the difference of attraction on two opposite points on Earth is greater for the Moon attraction than for that of the Sun. If we add 12,000 km to the previous calculation for the distance Earth-Moon and Earth-Sun, we find a gravitation difference of 9.5×10^{-7} m/s² for the Sun, and 2.0×10^{-6} m/s² for the Moon. The variations due to the Moon thus are 2.15 times more important than those due to the Sun, which explains why tides are more influenced by the Moon.

Problem 6.2: Why can't we see light bending?

We don't notice that light falls, because light goes much faster than usual objects. Its deviation is infinitesimal, but it exists on Earth. The scenario of Section 6.2.1 shows that during the time t needed for the beam to cover 10 km, the elevator has descended by gt. Hence, the beam is lower after 10 km by:

$9.81 \times 10{,}000/3 \times 10^8 = 3.27 \times 10^{-4}$ m $= 0.327$ mm.

Problem 6.3: Amazing effects when traveling in the cosmos: Bob and Alice Case Study 2

Q1: Bob will be younger than Alice because he inevitably incurs accelerations in his travel, whereas Alice remained fixed in a frame which is assumed to be inertial (cf. Section 3.5.5).

Q2: Bob's proper time on the massive planet is seen by Alice flowing more slowly than hers. Hence, this is another reason for Bob to be younger than Alice when he returns.

Q3: When Bob is at a higher altitude than Alice, and not close to another great mass, he sees Alice's proper time running more slowly than his own.

Q4A: For the same reason, Alice's voice will sound more to the bass.

Q4B: Bob's proper time is seen flowing more rapidly by Alice than her own; hence, his voice sounds more to the treble.

Problem 6.4: The clock in a geostationary satellite

Q1: The satellite being at a higher potential, its clock is seen running faster from Earth.

Q2A: The relation 6.3 gives the ratio: $\exp(\dfrac{\Phi_A - \Phi_B}{c^2})$, which is with the Newtonian gravitation:

$$\text{Exp}[\ \frac{GM}{c^2}\ (\frac{1}{6300}\ -\ \frac{1}{42300})] = \exp(5.99 \times x \times 10^{-10}) \approx 1 + 6 \times 10^{-10}.$$

During one day, which is 86,400 s, the satellite clock will be seen as having advanced by 52 μ.s.

Q2B: This duration of 52 μ.s represents $52{,}000 \times 30$ cm $= 15.6$ km.

Q3: The photons composing this signal acquire energy when falling to Earth; hence, their frequency increases, and so does the signal. This increase is by the same ratio as in the previous question; hence: $f_{eath} = f_s \times (1 + 6 \times 10^{-10})$.

Problem 6.5: Curved surface: the sum of the angles of a triangle

Q1: The angle $(AB, AN) = \pi/2$; the angle $(BA, BN) = \pi/2$. The sum of the angles of ABN is: $\dfrac{\pi}{2} + \dfrac{\pi}{2} + \alpha$, which is greater than π. Hence, the surface of the sphere is not Euclidean.

Q2: We have $S = \text{(sphere surface)} \times \frac{1}{2} \times \dfrac{\alpha}{2\pi} = 4\pi.R^2 . \dfrac{\alpha}{4\pi} = \alpha.R^2.$ Hence:

$$\alpha + \beta + \gamma = \pi + \frac{S}{R^2}.$$

Problem 6.6: Curved surface: Is the great circle the shortest path?

Q1: No, since the 3D geodesic is the shortest path, and it is the great circle.

Q2: The length of the path along the latitude is:

$$L = \varphi^\circ . \frac{\pi}{180} .R.\sin\theta = \frac{\pi}{2} . \frac{\sqrt{2}}{2} .R = \frac{\pi\sqrt{2}}{4} .R.$$

The length of the geodesic is: $l = R.\ \alpha$, with α being the angle (OA, OB) with O being the sphere center.

We have $\sin\left(\dfrac{\alpha}{2}\right) = \dfrac{AB}{2R} = \dfrac{\sqrt{2}R\sin\varphi}{2\times 2} = \dfrac{\sqrt{2}.\sqrt{2}}{4} = \dfrac{1}{2}$; hence: $\dfrac{\alpha}{2} = \dfrac{\pi}{6}$;

so: $\alpha = \dfrac{\pi}{3}$, then $l = \dfrac{\pi}{3}R$.

Finally, $l < L$ because $\dfrac{\sqrt{2}}{4} = 0.3535$, which is greater than 0.33333,

the difference being 5.7% ∎

Appendix 1: Summary of the Main Constants

A.1 Particles Mass, Energy and Charge

- $1eV = 1.602\ 176\ 634 \times 10^{-19}$ J. One eV is the kinetic energy acquired by one electron evolving in an electrical potential difference of 1 V during 1 second.
- Electron charge: $1.602\ 176\ 634 \times 10^{-19}$ coulomb.
- Atomic mass unit (abbreviation u): $1u = 1.66 \times 10^{-27}$ kg. It has been defined so that the mass of the most frequent isotope of carbon-12 is 12.00000 u. Besides, $1u = 0.9315$ GeV/c².
- Proton mass: 938.272 MeV/c² = 1.007276 u = 1.673×10^{-27} kg.
- Neutron mass: 939.565 MeV/c² = 1.008665 u = 1.675×10^{-27} kg.
- Electron mass: 0.511 MeV/c² = 5.49×10^{-4} u = 9.093×10^{-31} kg.
- Thus, the neutron and the proton masses are close to: 1 GeV/c² = 1.783×10^{-27} kg.

A.2 Main Constants

- Light speed: 299 792 458 m/s $\approx 3.\ 10^8$ m/s, or 30 cm/ns (1 ns = 10^{-9} second).
- Gravitational constant: $G = 6.674\ 30 \times 10^{-11}$ m³/kg s².
- Hubble time: $H_0^{-1} = 14.3$ billion years.
- Hubble constant: $H_0 \approx 70$ km/s.Mpc^{-1}.
- Reduced Planck's constant: $h/2\pi = 1.11 \times 10^{-34}$ J s.
- Planck time: 5.39×10^{-44} s.
- Planck distance: 1.62×10^{-35} m.
- Boltzmann constant: $1.38064852 \times 10^{-23}$ m² kg/s² K.

A.2.1 Earth

- Gravity on Earth: $g = 9.81$ m/s².
- Earth radius: 6,371 km.

- Earth mass: $5{,}972 \times 10^{24}$ kg (symbol: $1M \oplus$).
- Distance Earth-Sun $= 1.49\,597\,870\,700 \times 10^{11}$ m $= 1$ AU (~ 150 million km)
- Orbital speed of Earth around the Sun: $v = 2.98 \times 10^4$ m/s.
- Earth age: 4.57 Gyr (gigayears $= 10^9$ years, 1 billion years).

A.2.2 Sun

- Sun mass: 1.989×10^{30} kg (symbol: $1\odot$).
- Sun radius: 696,340 km.
- Sun luminosity $= 3.83 \times 10^{26}$ W (symbol: $1L\odot$).
- Sun age: 4.60 billion years.
- Gravity on Sun surface $= 274$ m/s^2.

A.2.3 Moon

- Moon mass: $7.34767309 \times 10^{22}$ kg.
- Distance Earth-Moon $= 384{,}400$ km.
- Gravity on the Moon $= 1.62$ m/s^2.

A.2.4 Interstellar Distances

- 1 AU : The astronomical unit (AU or au) is: $1.49\,597\,870\,700 \times 1011$ m $\sim$ 150 million km. It is the mean distance Earth - Sun.
- 1 pc $= 3.09 \times 1016$ m, or ≈ 3.26 light-years. It is the distance at which 1AU subtends an angle of 1 arcsecond.
- Distance to the nearest star, Proxima Centauri: 1.30 pc.
- Distance to the center of our galaxy (the Milky Way): 8,500 pc.
- Distance to the nearest galaxy, Andromeda: 0.76 Mpc (megaparsec).
- Universe observable size: 45.6 billion light-years, or $\approx 14\,$Gpc (gigaparsec).

***

End of Volume I
    Volume II *Advanced Topics* contains: complementary explanations, alternative demonstrations, further topics (notably in General Relativity and cosmology), additional case studies and exercises. Some parts require a higher mathematical level than Volume I, but the more complex notions and mathematical tools are carefully explained for those who do not have this higher background.

Bibliography

Abbott B.P. et al. (2016), *"Observation of Gravitational Waves from a Binary Black Hole Merger"*, Phy.Rev Lett. 116: 061102. doi:10.1103/PhysRevLett.116.061102

Bergmann, P.G, (1942) *"Introduction to the Theory of relativity"*, Dover publications

Blanco Laserna, D. (2014), « *Einstein et la relativité* », RBA France

Calle, C. (2005), *"Einstein for Dummies"*, Wiley Publishing, Inc

Clark, R.W. (1971), *"Einstein the Life and Times"*, The World Publishing Company

Cox, B., Forshaw, J., (2009), *"Why Does E = mc² ?"*, Da Capo Press

Damour, T. (2006), *"Once Upon Einstein"*, CRC Press, or in French: *"Si Einstein m'était conté…"*, Flammarion

Durandeau, J.P., Decamps, E.A. (2000) « *Mécanique relativiste* », Dunod

Einstein's (1912) *Manuscript on the Special Theory of Relativity*, a fac simile, G.Braziller publishers

Feynman, R.P, Leighton, R.B, Sands, M. (2011), « *The Feymann, Lectures on Physics* », Caltech

French, A.P. (1968), *"Special Relativity"*, WW Norton & Company

Galison, P. (2003), *"Einstein's Clocks, Poincaré's Maps - Empires of Time"*, W.W. Norton & Company

Halpern, P. (2016), « *Le dé d'Einstein et le chat de Schrödinger* », Dunod

Hawking, S. (1988), *"A brief history of time, from big bang to black holes"*, Bantam press

Hobson, M.P., Efstathiou, G.P., Lasenby, A.N. (2006) « *General Relativity. An introduction for physicists* », Cambridge University Press

Jungk, R. (1958), « *Plus clair que mille soleils* », Arthaud

Klein, E. (2016), « *Il était sept fois la révolution Albert Einstein et les autres…* », Flammarion

Klein, E. (2016), « *Le pays qu'habitait Einstein* », Actes Sud

Kogut, J.B. (2001), *"Introduction to Relativity"*, HAP

Le Bellac, M. (2015), « *Les Relativités: Espace, Temps, Gravitation* », edp sciences

Ougarov, V. (1974), « *Théorie de la Relativité restreinte* », Editions de Moscou

Pais, A. (2005), « *Albert Einstein La vie et l'œuvre* », Dunod

Penrose, R (2004), *"The Road to Reality"*, Vintage books

Resnick, R., Halliday, D. (1992), *"Basic Concepts in Relativity and Early Quantum Theory"*, Mac Millan

Rovelli, C. (2014), « *Et si le temps n'existait pas ?* », Dunod

Ryden, B (2017), *"Introduction to Cosmology"*, Cambridge University Press

Taylor, E.F, Wheeler, J.A. (1992), « *Spacetime Physics* », W.H. Freeman & Company

Thorne, K.S. (1994), « *Blacks Holes and Times Warps* », W.W. Norton & Company

Tipler, P.A., Mosca G. (2008), *"Physics for Scientists and Engineers"*, W.A. Freeman

Villain, L. (2015), « *Relativité restreinte* », De Boeck.

Index

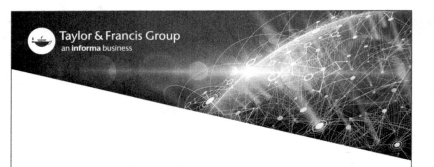

Printed in the United States
by Baker & Taylor Publisher Services